INVISIBLE STORM

ALSO BY JASON KANDER

Outside the Wire: Ten Lessons I've Learned in Everyday Courage

Courage Is
(coauthored with True Kander)

INVISIBLE STORM

A SOLDIER'S MEMOIR
OF POLITICS AND PTSD

JASON KANDER

MARINER BOOKS

Boston   New York

HarperCollins books may be purchased for educational, business, or sales promotional use. For information, please email the Special Markets Department at SPsales@harpercollins.com.

FIRST EDITION

Designed by Chloe Foster

Library of Congress Cataloging-in-Publication Data has been applied for.

ISBN 978-0-358-65896-2

22 23 24 25 26 LSC 10 9 8 7 6 5 4 3 2 1

To Diana.
My teammate, my soul mate, my best friend, and my hero.

men will literally run for president instead of going to therapy

—@Coll3enG

CONTENTS

PROLOGUE

On October 1, 2018, I walked into the Kansas City Veterans Affairs Medical Center and found my way to the small one-room office of a veteran service officer. Only two people in the world knew what I was doing that day: my wife, Diana, and my campaign manager, Abe Rakov.

I wrote my name on the sign-in sheet pinned to the wall outside and fell into line behind a handful of other vets, some young, some old, leaning against a wall in a hallway that doubled as a makeshift waiting room. All of us were new patients, waiting to be enrolled in the forbidding maze that is the VA system. Twenty minutes later, an overworked gentleman in a red American Legion polo shirt emerged from the office, glanced at the sheet, then looked up at me. His eyes widened.

"Whoa," he said.

"Yep. That's me," I replied.

I pulled my baseball cap down a little lower as I followed him into the tiny office. I was relieved when he shut the door.

I was hoping no one else recognized me. But there wasn't much chance of that. For most of 2016, you couldn't watch television anywhere in the state of Missouri without seeing my face at damn near every commercial break, either heroically lit up, smiling, and approving this message (those were my ads), or distorted and melting nightmarishly into the faces of Nancy Pelosi and Bernie Sanders (those were the other guy's).

My new friend began to go through the mental health intake questionnaire. Over and over, I found myself saying yes to his questions, and within minutes, he said, "It sounds like you need to see someone *today*."

The next thing I knew, he led me down to the emergency department and left me with a triage nurse, a very warm, older African American woman. She gave me a little slip of paper to fill out, and there were two questions on it:

"Have you had suicidal thoughts?"

"Yes," I wrote.

"Have you experienced intrusive dark thoughts? If yes, for how long?"

"Yes. Ten years."

The nurse looked at my form.

"Ten years?" she exclaimed.

I nodded.

"Honey, where you been!?"

The one place where you don't want to be famous is in a psych ward. As I went through intake, I caught staffers suppressing double takes when they recognized me. Finally I was put in a windowless cell with pale-green walls, where I sat hugging my knees in a hospital bed, dressed in a set of dark-green scrubs that were about five sizes too big. Because I'd said I was suicidal, the staff had taken all my belongings, including my belt and clothes, but I guess they figured that I wasn't going to kill myself with a paperback. They'd let me keep my book: an advance copy of Rick Ankiel's autobiography.

Unless you are a pretty serious baseball fan, you may not be familiar with Rick Ankiel. He was a big-league pitcher, a phenomenon, a major talent of his generation, until suddenly, in the middle of a playoff game, he lost the ability to throw strikes—forever. A friend in the publishing industry had randomly sent me the book. Now it seemed like a cosmic joke.

I wasn't really in the mood to read, though. A nurse was in there with me too, to keep an eye on me. When I had to pee, she turned her back to give me some privacy.

So this was suicide watch.

I sat there in my scrubs and I tried to wrap my head around where my life was headed. Answering yes to all those questions—*How often do you feel unsafe? Have you had thoughts of ending your life? Do you have difficulty sleeping due to memories of past events?*—had driven home just how desperately I needed help and how close I was to hurting myself, or worse. But I also knew clearly that just a few miles from here, a huge campaign machine was churning away to get me elected as the next mayor of Kansas City. I was going to win too—not only was I up in the polls, but I'd raised three times as much money as the nine other candidates *combined.*

It was funny, in a way—how many countless hours had I spent in windowless rooms, making calls for hours on end to ask campaign donors for money? At least this windowless room had a toilet, even if it was of the no-lid, stainless-steel variety.

Finally, after about half an hour, a psychiatrist, a young resident, came into the room. It was evident that he didn't know who I was, which was a huge relief. I was used to feeling people's eyes on me wherever I went, but usually it felt oddly comforting, even empowering. But here? It was humiliating.

For the next thirty minutes, I confessed everything I'd spent years hiding from the world: my night terrors, my consuming fear of someone hurting me and my family, my ever-present anger, my unrelenting guilt and punishing shame, my inability to feel joy, and my increasing dislike of myself. I told him how much of a burden I'd become to everyone around me. To my surprise, he seemed to take it all in stride. He asked what I had planned for the rest of the day, and I said I had to go pick up my son, True, from school at 4:30. "That's good," the doctor said, which I took to mean, "If you're making plans, you're not going to kill yourself *today.*"

But maybe he wanted to double-check. He reviewed his notes, looked up at me, and asked, "Do you have a particularly stressful job or something?"

Now, I was used to introducing myself dozens of times a day, but it hadn't been a real introduction for years. It was more like a pantomime of humility. When I said, "Hi, I'm Jason Kander and I'm running for . . . ," I was usually flanked by people wearing T-shirts with my name spelled out in giant letters.

So I said flatly, "I'm in politics."

He seemed curious. "What does that mean?"

I thought about listing off my résumé. Should I start with my time serving in the state legislature? Being elected secretary of state of Missouri? Running for the US Senate in 2016 and just barely losing? Should I talk about getting ready to run for president and giving speeches in forty-six states in the past year alone? Or how I decided to run for mayor instead? I figured I'd just cut to the punch line. "Well, I almost ran for president, but then decided to run for mayor instead, and tomorrow I'm planning on calling that off."

"You were going to run for president?" The doctor blinked a few times. "Of what?"

He looked confused. In fairness, you would be too, if you were a psych resident and some random thirty-seven-year-old in ill-fitting scrubs on suicide watch claimed to be a presidential candidate.

I knew it would sound silly to answer his question, but I did. "Of the United States."

He looked skeptical, or maybe he was suppressing a chuckle. "Who told you that you could run for president?"

At that point, I went from feeling mortified that everyone else had recognized me here to feeling irritated that this guy didn't believe me.

"I don't know what to tell you, man," I said. "I mean, I spent an hour and a half talking it over one-on-one with Obama in his office, and he seemed to think it was a pretty good idea."

The doctor sat back in his chair. "Barack Obama told you that you could run for president?" He tapped his notebook a couple of times with his pen, then pursed his lips. "So how often would you say you hear voices?"

1

THE UNIFORM

Something happens to you the first time you put on a uniform. I think this is true for anyone, from an orchestra conductor putting on a tux to a nurse slipping into scrubs to a UPS driver getting dressed in the brown ensemble: the first time you see yourself in that uniform, a strange clarity settles over you. Suddenly, you are part of something. When you look in the mirror, you don't see someone wearing clothes anxiously picked out that day. Instead, you see someone with a place in the world—creating music, healing people, delivering packages. A job. A purpose. Or in my case, a mission.

At 4:45 a.m. in Washington, DC, I put on an army uniform for the first time. It was 2002, and everyone else in our little apartment in Glover Park was fast asleep: Diana, my fiancée at the time, was in bed, and our rotund little pit bulls, Winston and Shelby, snored beside her. Stealthily, I pulled on the green-camouflage battle dress uniform I'd laid out on the dresser the night before. The pants, cinched with a standard black belt at the waist. The dark brown undershirt. The blouse—that's what it's called—with its sturdy buttons and extensive pockets. The fabric felt stiff. The texture and design of the uniform felt purposeful. It felt . . . *tactical*. It felt cool as hell. Just above the breast pocket was sewn, in bold black capital letters, U.S. ARMY.

I put on my camo cap and I checked myself out in the tiny porthole mirror over the sink in our cramped bathroom. Gone—or at least concealed—was the skinny first-year law student at George-town University, who was bored to tears by property law. Want-ing to get a better look without waking Diana or being seen doing something that still felt a little too much like playing dress-up, I tiptoed five steps into the living room and stood on the couch in my green socks. I leaned back, put one foot on a nearby table to get more height, and balanced warily as I looked into the gaudy mir-rored tiles affixed to a wall light that Diana and I had dug out of her parents' basement back home in Kansas City three weeks before. I could see myself only from the waist up.

From what little I could make out, I looked like a soldier. And the instant that thought crossed my mind, my mind shot right back: *No, you're not. You haven't done a damn thing to earn this uniform.*

Fair enough, I thought. But I would do so soon enough. I pulled on a pair of heavy black boots, haphazardly tucked the bottom of my trousers into the ankles, laced them up, and tucked the laces inside the boots. This rubbed up against my leg, and as I walked toward my red pickup truck, I felt a painful sting with each step. I made a men-tal note to find out how the heck the whole "boot blousing" thing was supposed to go.

It was my first day of ROTC.* The drive to the rear entrance of the student center on Georgetown's main campus was short. I saw a lot of uniforms darting about, counting people, saluting, and sounding off. In the predawn light and with my limited knowledge of insignia and nomenclature, I couldn't distinguish the returning cadets from the active-duty soldiers who made up the cadre. I spotted a hodge-podge group of cadets who looked especially confused. Some wore uniforms as pristine and unadorned—no patches or name tapes—as

---

* The US Army's Reserve Officers' Training Corps, or ROTC, is a four-year pro-gram leading to a commission as a second lieutenant in the army. It's typically com-pleted during college, but because I was a law student, I took a compressed version, three years instead of four. The first two years (for me, one year) are the equivalent of basic training; the third and fourth years are the equivalent of officer candidate school. You learn how to be a soldier in years one and two and how to lead soldiers in years three and four.

mine, while others were standing awkwardly and self-consciously in civilian clothes. This had to be my group: the newbies.

I filed past a kid in civvies who turned and bumped into me before straightening with fear and uttering, "Excuse me, sir."

I thought, *Dude, I'm as lost as you.*

The new cadets had been loosely organized into three parallel lines, so I planted myself at the tail end of the third. You hear army people talk about "falling in" all the time, but none of us actually knew what we were supposed to fall in to, or how.

We were marched—which took way more concentration than I'd imagined—into an auditorium for orientation briefings from cadets who had just returned from summer training at Fort Lewis, Washington. Two hours later, we were "dismissed" from something for the first time in our lives, and already an electricity was buzzing inside me, a flickering sense that I had begun walking a path I was born to follow.

As I walked back to my truck, I passed a cadet officer. Having learned just minutes earlier how to recognize rank insignia, I rendered my first salute, pulling back my shoulders and stiffening my spine, making a knife blade out of my slightly down-canted right hand until my index finger touched the brim of my cap, all without breaking stride. Just like the movie salutes I'd seen and imitated since childhood. The cadet officer returned this first salute of my career with a simple "Hooah," and continued past.

It felt incredible. I was so puffed up with pride, I could have floated away.

After I got back to our apartment and changed clothes, I was ready to let loose an absolute monologue of enthusiasm, but Diana was sitting on the couch surrounded by law textbooks and legal pads. She could sense that I was ready to burst, so she smiled, put down a textbook, and said, "Okay, I want to hear everything."

From the moment I'd met Diana, anything good that happened to me didn't feel real, didn't become real, until I'd told her all about it. The things I did mattered only if I was able to share them with her. We'd gone to rival high schools and met halfway through our senior year at a speech and debate tournament. She got up to speak and I was instantly transfixed. Growing up on the outer suburban edge of Kansas City, I'd never seen anyone quite like her. Shortly thereafter, we had our first date: my senior prom. She wore a black dress and a leather jacket, and when I picked her up, she said, "Feel my hair. It's basically a helmet." She was the least self-conscious person I'd ever met.

At first, it was her beauty that knocked me back, but by the end of that night it was her brain. She was smarter than me (and everyone else) yet never seemed interested in trying to impress.

Born in Ukraine, Diana came to the United States as a refugee from religious persecution at the age of eight. She had grown up poor and had been working at least twenty hours a week since middle school. My family, on the other hand, had been in Kansas City for over a century. Though I'd had a few summer jobs, I'd never needed to earn a single dollar. On paper, Diana and I had nothing in common, but somehow we had come to see the world in identical ways, including our place in it. We fell in love immediately and became inseparable.

Now that we lived together (and—lucky for me—had identical 1L course schedules), every night was a slumber party with my best friend. We'd stay up late, debating everything from American foreign policy to whether or not I could really eat all the food in the fridge in one sitting.

After I'd boiled down the entire two hours of morning briefings into a tight hour-and-a-half summation, I took a breath and asked her if I still had time to do the reading. "Of course not," she said with a laugh, "but don't worry, I took good notes and I'll tell you all about it on the drive to class."

DIANA

Jason and I met when we were seventeen. Our first date was his senior prom. I was an alternate date because the original one got caught smoking pot and was grounded, so she couldn't make it. She and I were on the same debate team. And as she and Jason were brainstorming options for his stand-in date, he threw out my name. Her reply was "Oh, definitely! Diana is totally nonthreatening."

We dated for four years, applied to and got into law school together, and were married after our first year, at the age of twenty-two. The thing we bonded over from that very first date and throughout our early relationship was how much we were both dedicated to changing the world for the better. We were going to do big things that had a positive impact on people's lives; we just didn't know how.

After Jason's first day of ROTC, it was pretty clear that it made him light up like nothing I had ever seen. How could someone enjoy having to get up so early, get yelled at, and learn about military tactics? For me, it held no appeal at all. But Jason didn't just enjoy it. He reveled in it. It was obvious right away that he had found his "thing."

The second day of ROTC brought me crashing back to earth, when, for my first-ever physical training session, I accidentally joined the Army Ten-Miler team. I showed up in the PT uniform (a less tactical-feeling gray shirt that said ARMY and black shorts), accessorized—in my case—by a giant plastic knee brace that ran from just above my ankle to halfway up my thigh. PT was where I knew I needed to prove myself and also where I felt the most intimidated.

Less than a year earlier, I'd planned out my ROTC path with an army recruiter, gotten into great shape, and then, while playing pickup football, promptly blew out my anterior cruciate ligament and meniscus. At the time, the army considered ACL reconstructions disqualifying, but there was a glimmer of hope. If, at the end of my first year of training, I had proved myself capable of keeping up, I'd be examined by an army doctor and potentially granted a waiver. Without that waiver, my military career would be done before it even began. For the past few months I'd ground it out through surgery and physical therapy, thinking about this moment every step of the way.

I saw a group of cadets running in formation, and I fell in—there's that phrase again—and started jogging along. A minute later I noticed we weren't running in step or singing cadence, and I felt a pang of disappointment—in movies, soldiers always sang cadence when they jogged. Then I realized why no one was singing: We weren't really jogging. No, this was running. We were definitely running. Fast. *Okay, this must be an initial sprint*, I thought, *just to jolt us awake, and then we'll slow down.* But we didn't slow down. Not even on the hills, and we seemed to be doing a tour of every steep incline on campus. I'd known PT would be tough, but within minutes my lungs were on fire and my knee was going, *Oh God, what are you doing?*

But I couldn't let on how challenged I was. Was I trying to prove the doctors wrong or convince myself I could do it? Not really. My motivation was more like social fear: I didn't want to fail in front of the other cadets, even though I didn't know anyone or recognize a single face from the prior morning. Fueled by peer pressure, I gritted my teeth and tried to push through.

After what felt like miles (it was probably more like one) we reached the Key Bridge and began to turn onto the towpath along the Potomac River. I gasped to the guy running next to me, who looked like he was out for a pleasant jog, "Do we always run this fast?"

He gave me a strange look. "I mean, on the Army Ten-Miler team, yeah."

A few minutes later, plastered with sweat, my knee the size of a grapefruit, I ran up to my sergeant and babbled my excuse for being so remarkably late. He cut off my little speech with a wave and a scowl: "Cadet, I'm familiar with the concept of lateness. I'd smoke you but PT is almost over, so just fall in over there, Hero."*

When I finally staggered back to my truck, mortified, my knee

---

* Bonus: I was also learning that laudatory-sounding monikers—such as "Hero"— were reserved exclusively for sarcastic purposes.

feeling like it had been carved open, all I could think was *That sergeant is scary as hell.*

And I want to be just like him.

A lot of politicians like to say they didn't plan to go into politics. Sometimes they're even telling the truth. I definitely planned on doing so, or at least I did when it finally became clear in high school that I wasn't destined for professional baseball. When I was fifteen, I participated in a Young Leader jamboree-type trip to Washington, DC. Each of us could choose one tour: the White House or the National Archives. I opted for the archives, proclaiming, with true teenage arrogance, "I'm not going into the White House until I live there."*

Joining the military, on the other hand, hadn't been part of my plan. It's hard to remember now, but at the turn of the millennium, the United States had gone nearly an entire decade without engaging anyone in a full-scale war. I thought it would be cool to be a soldier, sure, and I had a vague notion of it as a résumé enhancer, but I'd had only cursory visits with recruiters during high school and didn't seriously consider joining the ROTC when I started undergrad. Diana was interested in service too. In high school she filled out a contact form with an army recruiter, but both times when the recruiter phoned the house, her mother answered and assured them they had the wrong number. In college, Diana became really interested in the FBI. She must have read ten books on its history and how best to become an agent. The desire to serve was something we bonded over, but in my mind it still lived in a "maybe someday" place, as in "Maybe someday after I finish law school, I'll join the Judge Advocate General's Corps as a reservist or something."

I've always had a deeply protective streak. I get this from my parents, Janet and Steve, who met when they were juvenile probation officers. Dad told us how Mom had single-handedly disarmed knife-

---

* Of all the candid admissions in this book, this is possibly the most personally cringeworthy.

wielding seventeen-year-olds back when Mom and Dad's dates consisted of serving juvenile warrants together, and this wasn't hard to imagine. Throughout my childhood, our family took in neighborhood kids who were having trouble at home. My younger brother, Jeff, and I didn't have to ask our parents to invite our friend Mel or Justin or Dan into our home. They just did it. And in our home, one rule united us all: you protect your people.

Dad often told me how much he regretted not joining the military. In the 1970s, he had been a civilian aviation instructor for the air force, training young pilots in Texas, and he had been offered a commission to fly in the Air National Guard. He took a pass, but when I was growing up, he'd muse a couple of times a year about what rank he'd have reached by that point, had he made a different decision. A C-130 would fly over the house on a weekend, and Dad would say something like "St. Joe unit getting their hours in. I might've been their squadron commander by now."

During World War II, Dad's dad, my grandfather Ed ("Pop," to my generation), had flown all over North Africa as a radio man in the Army Air Corps, though he rarely spoke about it. It wasn't (to my knowledge) that he'd seen anything particularly haunting. It was the opposite: Pop didn't think his experience was special. There was a war on, so he went, like everyone else he knew. Just like his father, my great-grandfather, had done in World War I.

The fact that they had joined the military didn't seem crazy or dangerous to me—it seemed natural. It was what people did when there was a war: you went and fought, and then you came back and lived your life. But you *went.* A part of me felt envious of my grandfather and my great-grandfather.

Then the planes flew into the towers and the war was here. *My* war.

On 9/11, I was in my final year of undergrad at American University. By the afternoon I was standing in line to give blood a few blocks from the Capitol. After a long wait, a woman came out and said they didn't have the capacity to take more donations. "I hope you can find some other way to help," she said. I knew then and there that I was going to join the military. In the weeks that followed, I researched my options, spoke to recruiters, and started doing push-ups.

My knee injury prevented me from enlisting right away. But I found out that I could become an ROTC cadet when I started law school the following year and could enlist in the Army National Guard after my first year of training.*

When I told my professors at American University about my plans, they looked as if I'd told them I was dropping out to follow a passion for ditch digging. It didn't make sense to them. And at first, I saw where they were coming from. I was at a private East Coast university and headed to a private East Coast law school, and hardly anyone there knew someone in the military. But back in Kansas City, I knew two or three guys from my high school baseball team alone who'd joined up even before 9/11. Where I grew up, most everyone viewed the military as a solid option after high school, on a par with college. But where my professors grew up, the possibility of joining up was a cautionary tale, something that might happen if you couldn't get into college.

Additionally, most of my professors had come of age during the Vietnam War, when dread of the draft was part of everyday life. So after my knee surgery, they looked at my crutches and saw an uncashed lottery ticket. To them, my choice to enlist anyway made no sense.

I understood their point of view, but after the hundredth time a classmate said, "But you have an education," or a professor told me military service would be a "waste of someone like you," their attitude began to upset me. Heck, even members of my extended family had implored me to find "better ways to serve." I was proud

---

* A lot of ROTC cadets enlist in the Army Reserve or Army National Guard once they've completed the basic training part of ROTC. It's called the Simultaneous Membership Program, which means you're both a reservist and an ROTC cadet at the same time.

of my decision but also scared about what I was getting myself into; having to constantly explain myself took its toll. Eventually things came to a head.

One of my uncles and a few of his work colleagues were in Washington, DC, for a business meeting with Dr. Patch Adams, the physician who just a few years earlier had been portrayed by Robin Williams on the big screen. My uncle thought I might like to attend his pitch meeting to meet Dr. Adams, and he was right, so when he called and invited me, I didn't hesitate.

I sat quietly and enjoyed the show as Dr. Adams held court over drinks at a DC bar; my uncle and his colleagues were spellbound. Adams was gregarious and more than a healthy fraction as funny as the Robin Williams version of himself. After thirty minutes or so, my uncle charitably tried to include me in the conversation, telling the assembled group that "young Jason here is in the process of joining the army."

Dr. Adams had a strong reaction to this news. "Well, that's a terrible idea," he said.

I'd never heard that response before, and I was stunned into silence.

Sensing he'd dampened the mood at the table, Dr. Adams tried to lighten things up with a little comedy, and I should've gone with him, but our exchange had put me on the defensive. When he stood up to demonstrate how he'd gotten out of serving in Vietnam by pretending to be gay for the draft board, everyone cracked up—except me. I realized then what made me so mad about this and just about every conversation I'd had since deciding to join: the assumption that the military was for "other people," those who weren't special, who weren't white, moneyed college kids, who weren't like us.

I was too angry to let it go. I asked, "Do you ever wonder if the kid who had to go in your place got killed?"

It was like a record had screeched to a halt. The laughter evaporated.

Vaguely I knew I shouldn't have said anything. Dr. Adams's story was just a thoughtless spur-of-the-moment anecdote. Who knew if it was even true? Maybe someone else in that situation would have seen the bigger picture—I was a guest of my uncle, who was there trying to do his job; Dr Adams was a respected, beloved figure. But I couldn't let the moment pass.

Dr. Adams was graceful about it, changed the subject, and chatted for an acceptable period of time before excusing himself. I later felt bad and wondered if I'd blown a professional opportunity for my uncle, who was generous and forgiving about the situation. But it wouldn't be the last time I got a negative reaction when I told someone I'd signed up.

Eventually I started giving the same one-line response when someone questioned my decision: "If I can be good at the job, maybe some other people get home safely." It was a corny sentiment, and the look people gave me when I said it confirmed this. Yet it genuinely summed up how I felt.

I hurled myself wholeheartedly into ROTC training. And very quickly, I realized that I loved it—maybe a little too much. I was a first-year law student who was supposed to be poring over the mysteries of legal theory, outlining cases and figuring out which note-taking method, highlighter or sticky note, suited me better. But I could not have cared less. My desk was covered with manuals on small unit tactics, troop-leading procedures, counterinsurgency, anything I could get my hands on. In my first couple of semesters, I felt genuine enthusiasm for the assigned law texts maybe three times, whereas I inhaled and fully memorized *Field Manual 7-8: The Infantry Rifle Platoon and Squad.* While my law school classmates were spending weekends in the library, worrying about what might be on the final exam, my "study group" and I were bused to nearby army bases to learn how to conduct reconnaissance, assault an objective, and fortify a position. And when I got home, I'd collapse into bed, punch-drunk and happy, and Diana would help me untie and tug off my boots as I told her everything I'd done.

As supportive as she was, though, Diana worried a lot about what I'd gotten myself into. She'd been concerned since 9/11 when

I'd told her I planned to enlist. Mostly she kept this unease to herself—and we sort of silently agreed not to talk about a future deployment.

Halfway through the second semester of my first year of law school, an army doctor cleared me for full duty with no limitations and—in addition to my ROTC training—I was scheduled to enlist in the Maryland Army National Guard as a member of the 115th Infantry, a unit that had landed at Normandy. With two weeks to go before final exams, my Georgetown Law classmates were stress-eating and outlining recordings of lectures they'd already attended in person. Meanwhile, I was at Fort Belvoir, in northern Virginia. For four days, we Hoya Battalion cadets had been living in the woods. I hadn't slept at all, hadn't showered, and had been eating nothing but MREs (Meals, Ready-to-Eat). It was my first simulated combat exercise, with active-duty soldiers posing as insurgents. We moved round-the-clock, rain or shine, conducting patrols, receiving and executing a variety of missions during the day, and then digging our holes and sleeping in shifts at night.

When the exercise ended, we emerged from the woods and marched for miles toward the buses, lugging a good fifty pounds of gear and ammo on our backs, our faces smeared with mud and camo paint as we sang cadences like "Gory, Gory, What a Hell of a Way to Die!" A third-year cadet who had been my platoon leader all weekend came alongside me. Dog-ass tired, I looked at him, and he looked back at me with the biggest grin.

"How fucking great is this, Kander? We did some real *army* stuff today!"

He was right. It *was* great. We laughed the maniacal laugh reserved for two exhausted people who know something no one outside their tribe can know, and on we marched, singing about dying at the top of our lungs.

This, to me at the time, was the truest test of manhood—whatever that meant. I wasn't entirely sure. It wasn't like I'd grown up with a sense that being a man meant being tough or flinty or eating a lot of meat. Even though my parents *were* tough. My dad was a former cop who fully recovered from breaking his back in a plane crash and never counted calories or even shaved his mustache (I've literally never seen his upper lip), and my mom grew up in foster care and put herself through college. Still, they didn't want their sons to commit to the leathery Clint Eastwood archetype. I knew that being a man meant being dependable, taking care of your people, and going where you're needed—and I was desperate for anything that would let me be that man. In college, I'd write long emails to Diana, loaded with agonized questions about whether I'd ever done anything that mattered. "How can you ever be a real man," I'd ask, all of nineteen years old, "if you've never been tested?" I'd pledged a frat, and their whole initiation deal was forcing us to walk everywhere. No buses, no Metro. And I remember thinking, *This is fucking stupid. This is nothing.*

But this training? This was a test. This was real army stuff. I was exhausted, filthy, racked with pain from my feet to my ears, and I'd never been so damn happy. My whole life I'd wanted to be a part of something that would demand everything I had. And now I had that.

In the two years that followed, between ROTC, summer training at Fort Lewis, and serving with my national guard unit, I became increasingly comfortable with and adept at the ins and outs of soldiering. The learning had been so gradual and difficult that I'd hardly noticed just how full my tool kit had become. I knew how many of my strides it took to cover a hundred meters, and I knew how much to increase the number of strides when I was in the woods. I could hike like a mountain goat, sleep on any surface, and—given a ten-digit grid coordinate, a map, and a compass—I could find you a penny in the deepest woods. And if you took my compass and left me the map, I could find my way home by using terrain features. I could pack any garment into any size space if I rolled it tightly enough. I was passionate about foot care, evangelical about the importance of clean socks, and able to treat blisters better than a lifelong podiatrist could.

I understood which part of my foot needed to contact the ground first so I could walk silently on dead leaves, and I could communicate every basic small-unit command tactic with hand signals alone. I could recognize many maladies and injuries, from heatstroke to a sucking chest wound; I could carry a man twice my size, apply a tourniquet, and start an IV. I could clear a jammed rifle in seconds, and I could clean one in the dark. I knew so many army acronyms and slang terms that conversations with my new friends sounded like they were conducted in a foreign language.

In short, I had acquired the skill set of the average American soldier.

On May 20, 2005, three years after first donning a uniform, I received my commission as a second lieutenant. It was the day before Diana and I—now married—graduated from Georgetown Law, and both of our families had come to DC. I didn't particularly care about graduating, but I was proud as hell of becoming an officer. From the moment I entered law school, I'd known I would finish because they'd keep cashing my tuition payments—and frankly because it just wasn't that hard. That was different from army training. Plenty of times—like when I'd been up for two days straight on an exercise and couldn't focus enough to properly call for artillery support—I doubted whether I'd earn a commission.

At the ceremony my family saw me in uniform for the first time—I'd never had cause to wear one around them before. My national guard unit was in Maryland, and trainings were all over the place. I had yet to do anything of note, so the front of my Class A's were adorned only with my nameplate, a U.S. on each collar, and a gold dagger on each lapel, signifying my assignment to the Military Intelligence Corps. To me, it wasn't yet much to look at, but I will never, so long as I live, forget the look on my dad's face when he saw me.

When I greeted everyone—Mom, Dad, my brothers, my grandparents, and my in-laws—at the entrance of Georgetown's iconic Healy Hall, Dad literally stopped in his tracks halfway up the steps. For the entire morning, I noticed him gazing at me with pride whenever I wasn't looking. That alone made me feel ten feet tall.

As for my grandfather, I'd discovered that he'd earned two commendations during World War II, but he'd never received them. So I had a ribbon rack prepared, and he pinned it to his blazer. It was the first time I ever saw Pop display pride in his service.

And pride is exactly what I felt sloshing warmly around inside my very soul as I raised my right hand and took the oath, shoulder to shoulder with people I hadn't known three years before but now counted among my closest friends. We stood at attention, eyes to the front, so unable to contain our mutual joy and relief that as the band played the notes of "The Army Song," we could be heard too, humming the tune in unison.

The army didn't just give me meaning; it gave me a sense of order. At last I knew what I was doing and what I was doing it for. I knew what clothes I was wearing, I knew who my boss was, and more than that, I knew what my *mission* was. I'd found what I wanted to do in life. The word "cadet"—which, more often than not, I'd heard used derisively, like a slur—was finally behind me. Pinned to my shoulders in its place was the gold bar of a second lieutenant. I wasn't an imposter. I had the real uniform.

Now it was time to earn it.

2

AN ORDINARY DAY
(IN JALALABAD)

I've often been asked, "What was Afghanistan like?" It's a simple question for which I've never been able to muster a simple answer. If you want people to understand a war, you don't just tell them about the biggest events or the scariest moments. Because war is not only smoke and fire, it's the voltage of danger that seems to hum just under the surface of the whole terrain, like the sound of a fridge at night. So here is one of those ordinary days, one of a hundred or so. I picked it mostly because I remember how much I liked the weather . . .

As dawn light poked through the window of the little wooden hut, I could feel the warmth on my eyelids but chose to lie still and rest a little longer. I considered opening my eyes when I heard Tommy* coming to life in the bunk beneath me, but I held still. How much was it going to hurt to move?

   For the past two days, my lower back had been jagged with pain, which wasn't so much a concern as the stiffness and lack of mobility it was causing. The pain had begun as we'd made the three-hour trip from Kabul to Nangarhar through the treacherous J-Bad

* Here, as in certain other cases in this book, I've changed a person's name in order to protect identity.

Pass, our driver's side tires just inches away from the edges of thousand-foot cliffs and our passenger windows offering a view up the mountain, where expressionless men eyed us from their perfect perches above.

As we made our way around Jalalabad, the pain got even worse, as if someone had clamped a wrench around my lumbar vertebrae and was slowly twisting them. I'd become keenly aware that if we were attacked, I'd have a hell of a time escaping the back seat of the Chevy Suburban. On the other hand, I felt fortunate. For only the fourth time in the dozens of days I'd spent outside the wire, I was on the inside of something armored.

The soundtrack of Tommy's boots hitting the gravel outside convinced me to grit my teeth and sit up. *Holy shit!* I thought. My back didn't hurt. The relief was enormous. By some miracle, at this remote forward operating base I had awoken completely pain-free. It was like reaching into my pocket and finding an unexpected hundred-dollar bill.

I reached toward the foot of my bunk, picked up my boots, turned them upside-down, and shook them. Scorpion-free.

Today was going to be a good day.

I had slept in an olive-drab undershirt and the tan cargo pants I'd borrowed from one of my roommates back at my safe house in Kabul. It was December 2006, winter in Afghanistan, and though the climate east of the pass was much warmer, it was still frosty in the morning here, and I was thankful to be doing the type of work that offered me one more sunrise without having to shave with icy water, as I did on the days I wore a uniform. For the work I was doing today, not only were beards actually encouraged, but I could throw on a dark-gray baggy fleece with GANGES MARINA embroidered on the left chest. I'd bought it during a family vacation in the Pacific Northwest years earlier, and it probably never expected to cross an ocean in a duffel bag and then suffocate under body armor.

I fastened my holster to my hip and secured the lower strap around my thigh, then pulled my rifle and my pistol down from the top bunk where I'd been sleeping. I inserted a magazine into each and started to think about breakfast.

There was no mirror to look at, no smartphone to take a selfie. But I didn't need a mirror to tell me how I looked. Something incredible happens in a combat zone when you don't have to put on a uniform: you transcend being a mere soldier. You become a cowboy.

Most soldiers in Afghanistan came as part of a unit, flying over together, crammed into a giant C-130 Hercules alongside the men and women they'd be working with and fighting beside throughout their tour. I had come alone.

My job in Afghanistan wasn't exactly what I'd been trained for. In fact, it was several levels above the training I'd received at intelligence school. It was like going from college baseball directly to the majors, and there's a reason why almost no one does that.

By 2006, we had been at war in Afghanistan for five years—longer than the United States had fought in World War II. The Taliban exercised battlefield authority over large swaths of the country, and governed some too; along with Al Qaeda and other Pashtun militias, it conducted frequent terrorist campaigns in Kabul. The overwhelming majority of Afghans wanted the US-led coalition and the still newly formed government of Afghanistan to succeed, but few believed either could. The coalition's job wasn't a matter of gaining the public's favor but rather gaining its confidence. The government of Afghanistan had become the center of gravity in the war. If we could keep it from exploding, the population would come to believe that victory over the Taliban and Al Qaeda was possible, and—so the theory went—would openly join our cause. This, broadly, was the mission.

The US ambassador to Afghanistan and the general in charge of Combined Forces Command–Afghanistan were interacting regularly with high-level members of the Afghan government. Likewise, their subordinate commanders were doing the same with provincial governors, district governors, Afghan kandak (battalion) commanders, police and border officials, warlords, and so on. All of this presented a glaring problem. To be effective in these inter-actions, these officials needed detailed knowledge about the "extra-curricular activities" of the Afghan people they were dealing with. Basically, how corrupt, how much in bed with the enemy, how en-meshed in narco-trafficking were these folks? And even if they were corrupt—and they always were, to a lesser or greater extent—were they still competent? Competent, reliable officials were worth their weight in gold, so everyone was willing to overlook a little corrup-tion if it meant we could do business with them.

The person responsible for distributing these top-secret written assessments to the general and the ambassador, and indirectly to everyone else down the chain of command throughout the country, was my boss, Colonel Jack McCracken,* commander of the Joint In-telligence Operations Center (JIOC). Colonel McCracken served as the J2, meaning the director of all US intelligence in Afghanistan.

When I first arrived, Colonel McCracken told me he needed to fill two spots: one for an intelligence analyst on the night shift to summarize and analyze intelligence after someone else had taken risks to collect it, and the other an entirely new role. The colonel needed to send all this gouge (slang for intelligence) up the chain, but he didn't have anyone primarily assigned to keep track of who the hell these Afghan officials were, and what the hell they were up to. We weren't just working with them; we were, in many cases, re-lying on them for assistance, intelligence, and even security, which meant that the turbulence inside the Afghan government was just as dangerous as the chaos outside it. Colonel McCracken informally referred to the new duty position as "the internal stability person," which sounded pretty optimistic, given the situation on the ground. But unlike the analyst job, this one would require someone to go out and actually *get* the information that would then be analyzed. I didn't hesitate: internal stability guy was obviously the cooler and more important job. I asked if I could have it.

I was only a second lieutenant, and I had just replaced a lieutenant

---

* I'll bet you think that's one of the names I changed, but you'd be wrong.

colonel,* but I was the only new intelligence officer Colonel Mc-Cracken was going to get for a while, so he said, "Well, you have a law degree, so that's something." The job, obviously, had nothing to do with having a law degree. Didn't matter. Boom. Hired. He decided to call the position political-military intelligence officer. Any information gaps in the "pol-mil" area would be referred to me.

The job came with a friend. Salam, a brilliant and good-humored Afghan American man in his fifties, was assigned as my translator. Born in the United States but raised in Afghanistan, he returned to the States after graduating from high school. He signed up to serve as a translator immediately after 9/11 and had already been in country five years when we met. Unbelievably, Salam's hometown was Kansas City.

In a sense, I was now a gossip columnist in Kevlar. I regularly attended two or three set meetings where I would sit down with intelligence officials in the Afghan army and let them brag about foiling attacks on US forces or listen to their theories about which supposedly friendly warlords aided such attacks. I'd make note of these details, go back to base, and write them up. Rarely did I analyze any of this information; I'd just put it into the classified system so it was available to those needing it. These meetings were fascinating (and not very hazardous), but they didn't take up even 20 percent of my time.

Mostly I drove around Kabul and the surrounding area with Salam, just a couple of Royals fans bebopping around a war zone, making contacts, talking to them, letting them dish, and getting them to introduce us to other Afghans who could do the same.

In these meetings I got the stuff my Afghan counterparts either didn't know or didn't dare pass along. It ran the gamut. On a given day I might learn about a high-ranking official in bed with Al Qaeda, a provincial governor running protection for traffickers along the heroin highway, a Taliban commander looking for an opportunity to

---

* There are four levels between those ranks, or in other words, about twenty years of service.

switch sides, or an Afghan army officer who refused to stop raping his soldiers. I heard stories of espionage, incompetence, and outright insanity.

Day in, day out, Salam and I would hit the road in our anonymous midsize SUV, looking for all the world like two regular Kabul dudes in a janky Mitsubishi. This was the bulk of my job—leaving the safety of the base to collect intel. Outside the wire, we weren't on comms, we weren't being tracked; we were almost entirely free-range. Often I'd be gone for an entire day, and no one would know where I was.*

When it wasn't just Salam and me, it was me and the tactical human intelligence team (THT), Todd and Kevin, more experienced intelligence officers who had actually been trained in all this secret squirrel stuff that I was learning on the job. They both outranked me, but they wore beards and street clothes full-time and went by first name instead of rank.

I learned a lot from Todd and Kevin, and because they were the only other guys on our little base doing this kind of work, the four of us—including Salam—grew close. Todd was a big, burly father figure who made everyone around him feel safe. He didn't say much, but he seemed fearless. Kevin was more intellectual. He didn't just wear street clothes, he wore distinctly Afghan street clothes. He fasted during Ramadan and often joined in with the interpreters during daily prayer. Todd—like most of us—cared about the Afghans, but Kevin loved them.

It always made me laugh that Kevin, despite his efforts to assimilate, couldn't break the habit of using his turn signal when driving. This drove Todd crazy. "Dude, nobody over here does that,"

---

* The many different disciplines of military intelligence are referred to with abbreviations, generally ending in INT. For instance, monitoring communications is SIGINT, or signals intelligence, and managing human sources is HUMINT, or human intelligence. What Salam and I were doing was probably closest to HUMINT but didn't fit the category perfectly. Years later, Colonel McCracken would tell me he thought of the work we did over there as "THUGINT," meaning we built relationships with thugs so that they'd give us dirt on other thugs.

he'd say. "We may as well put a Florida license plate on this thing."* Todd preferred to do the driving, and it always impressed me that he seemed to know Kabul—including all its little side streets—like the back of his hand.

By the end of the day, whether I'd flown solo or rolled with the THT, I usually had several morsels of information worthy of being written up as intelligence. I'd spend a few minutes at my desk, entering a couple of paragraphs into the combined US intelligence database for Afghanistan. This enabled unit intelligence officers across Afghanistan and analysts back in Qatar or Tampa to do keyword searches and then read what I had just heard some dude say he heard from some other guy.

In between these tasks, I volunteered for and eventually commanded convoys to and from places like Bagram Air Base. I also filled in for Colonel McCracken at the occasional high-level meeting (let me tell you, Pakistani colonels *love* being asked to meet with a twenty-five-year-old second lieutenant), augmented the THT on other missions, and vetted requests by members of the coalition to disclose intelligence to Afghan counterparts.

Among these extracurricular projects, I'd forged a working relationship between the JIOC (my unit) and the psychological operations team working out of Kabul. This was how I, underslept and starving, ended up in Jalalabad on this particular morning.

If you ask people their favorite type of food, they might say tacos or barbecue or Chinese, but for my money, nothing beats Army Breakfast. Chock-full of starchy carbohydrates, animal proteins, and calorie-rich saturated fats, it may have been engineered to be cheap or just to power someone in a physically strenuous job through to their next meal—I don't know. But either way, mission accomplished.

Scrambled eggs, hash browns, grits, cottage cheese, bacon, sau-

---

* Todd had a dry, sardonic wit. Once, during a USO show at Camp Phoenix, he turned to me and said: "I just came from investigating a suicide bombing where I took pictures of the dude's severed head, and now I'm at a Darryl Worley concert. War is fucking weird, man."

sage patties, and—in my case—more cottage cheese—all doused in Tabasco. Objectively, the army version of these items may not be better than their counterparts in civilian life, but processed food suits my Midwestern palate. And you can't beat the portions.

The breakfast at this forward operating base (FOB) was subpar by army standards, but it's unfair to expect much from a chow hall with so few mouths to feed. Plus, I couldn't really complain about runny cottage cheese and spongy hash browns when I was looking out at a gorgeous view: poppy fields sweeping out toward snow-pinnacled mountains.

The place was mostly empty except for a couple of contractors serving the food, a few random pairs of soldiers who lived on the FOB, and the motley crew that Tommy and I had cobbled together for this particular four-day "drug deal"* of a mission.

I set my tray down, but before I dug in, I paused to pray over my food for a moment. I had never done this before my deployment, but lately it had become a habit. It wasn't just because I was in a dangerous place—that was only a small piece of it. By this point I'd been eating meals surrounded by soldiers for three years, and it wasn't uncommon for my tablemates to offer a silent prayer before eating. I was often the only Jewish guy in my unit, and it wasn't like my Jewish education had been that robust—I knew hardly any Hebrew prayers.† But in Afghanistan, everyone did this, including Salam, my most frequent tablemate. I spent most every day so overcome with gratitude for the fact that I was actually over there, doing something that seemed so important, that I just felt really thankful. And so, before every meal, I would bow my head, close my eyes, and thank God for the chance to serve my country.

I know how this sounds. If I read that line in a book, I'd think, *No*

---

* I'm not sure why any negotiated arrangement is called a "drug deal" in the military, but I've always found the term charming. Being surrounded by poppy fields only made it more apt.

† If you want to fix me on the Judaic spectrum, I never had a bar mitzvah, but I stomped the glass at my wedding.

fucking way, but I'm telling you the truth. I just felt so blessed by this experience. And I probably figured that if God had an eye on me at all, the constant gratitude couldn't hurt my chances of staying alive.

I sent my prayer on its way, dug into breakfast, and began to go over the day's plan with Tommy.

There were nine of us on the op: two from intelligence, three from psychological operations (PSYOPs), and a four-man mobile training team (MTT). Aside from the convoys I ran, this squad-sized element was by far the largest crew I'd rolled with in country, and I was enjoying the camaraderie.

Tommy was the senior psychological operations guy. PSYOP teams were made up of soldiers from Special Operations Command who had been specially trained to "influence the emotions, motives, reasoning, and ultimately the behavior of governments, organizations, groups, and individuals." Basically, PSYOPs finds ways to "win the hearts and minds" of the population, and when that doesn't work, the teams amplify the appearance of strength to deter the population from helping the bad guys.

In order to accomplish this, PSYOP teams—like some other spec ops units—enjoy a high degree of independence to choose, plan, and execute their own missions. As such, they operate in small numbers and provide their own security, so their training includes a lot more than just mind games, or as Tommy explained it to me when we'd met a few weeks earlier, "We self-protect." That idea resonated a lot with me.

Tommy was about my age. He was a tall, slim Black man with a close-cropped army haircut, a well-kept beard, and a soft-spoken confidence that I found reassuring. He was a staff sergeant, and under normal circumstances, that's how I'd have referred to and addressed him, just as he normally would've addressed me as lieutenant or sir. But when you're rolling in street clothes so as not to appear military, it also makes sense to avoid sounding military.

For this mission, our vehicle carried three other team members. Ben, a blond, baby-faced twenty-year-old corporal from Springfield, MO, had achieved the out-of-regulation "operator" look by growing out his hair in a mop cut. With his M-4 rifle at the low-ready he looked like one of the bank-robbing surfers in *Point Break*. The other was Farid, our translator, who was a little younger than Salam (who'd stayed in Kabul this time) and had more of a soldier's mentality than the average translator did. Clearly, Farid had received substantially more training in weapons and tactics; Tommy and Ben trusted him to do a lot more than translate. Finally, Airman Cook, who was just called Ellen during this trip, was an analyst at the JIOC who could tell the PSYOP guys everything they needed to know about Jalalabad's power structures, warlord rivalries, and local terrorist leaders, even though this was her first time actually setting foot in what we referred to as RC-East (Regional Command–East). Like most female service members in country, she wore a scarf to cover her hair when she was dressed in street clothes. She joked that it went well with her rifle.

Today we would be rolling with the MTT, which was made up of three comparatively high-ranking US marines and a short, gregarious sergeant major from the Afghan National Army (ANA). He was fluent in broken English and insisted we address him as Rambo. These four guys usually wore uniforms and rode in a separate vehicle; they didn't join us on every excursion.

We had three missions planned for the day: meet with a local man who might give us information about an Afghan official I was investigating; evaluate and begin training a fresh group of ANA recruits; and use those new Afghan trainees to do a little recruitment show for the public. The meeting served my intelligence goals, and the training and displaying of the incoming Afghan soldiers met the needs of the MTT and the PSYOP crew.

I was technically the third-ranking person of the nine, but out here rank was immaterial; we followed the lead of whichever element was in charge at a given moment. So, depending on the circumstance, it was the marine lieutenant colonel, me, or Tommy who handed out orders. And when I was in command, I was pretty open to Tommy's suggestions.

After we inhaled as much breakfast as we could, Rambo and the marines headed to a hidden remote site where the rest of us promised to join them later. Our vehicle was bound for the office of a local "businessman" none of us had ever met. The purpose was intelligence collection, and Tommy and Ben would be providing security, which was a luxury for me. Salam and I usually had to roll without it.

We chambered rounds into our already loaded weapons, mounted the Suburban, and left the relative safety of the FOB. Tommy was driving, and in the left bucket seat, right behind the driver and secured with a seat belt, was a large potted plant. A warlord had gifted it to us the previous day, and we didn't want to seem ungrateful if we ran across one of his acolytes during our travels. Plus, it had sprouted some yellow buds, and we were curious about what its blossoms would look like. Farid had taken on the responsibility of watering it.

Twenty minutes later, we arrived at a two-story office building several blocks from the more bustling parts of Jalalabad. It was still early morning, so Tommy was able to back the Suburban into an adjacent alley and park beneath an overhang. At the door an Afghan man in his late sixties or perhaps his early seventies, with a brown-and-gray beard and the typical "Manjamas,"* met me, Ellen, and Farid. For just a second his gaze paused on Ellen, but he had evidently done enough business with the coalition to contain his surprise at the presence of a woman. He did look a little curious about the plant but didn't say anything.

As he led us inside and up a flight of stairs, I had just enough time to peek at the first floor, to see what was there. The answer was nothing—it was just an empty room with a front-facing window and no other interior or exterior doors. No bathroom, not even a closet. The wooden stairwell was dark and enclosed; it led to a small office.

The man took a seat behind a large wooden desk beneath a window that looked out on a small rear parking lot. Shrubby trees and power lines obscured the view. The room was wood paneled and carpeted, but like all interiors in Afghanistan, its surfaces were shrouded in

---

* Our admittedly insensitive term for the loose, flowing white shirt and pants and oversized open vest that were popular among Afghan men.

fine dust. The "businessman" gestured for us to join him. When he offered me a chair facing the desk, I obliged, though it meant I had no view of the pair of closed doors behind me. Willingness to take this somewhat vulnerable position was offered as a gesture of trust, which helps when you want people to tell you things they shouldn't. But every time I did something like this, I felt like someone was about to throw a bag over my head and stuff me into the trunk of a car. That was the irony of this pantomime of trust—it just reminded me of how little I actually trusted anyone here. Ellen sat in the chair to my left, and Farid had settled into a couch to the old man's right, facing all of us.

The walls were bare. No phone, collectibles, ashtray, or even paper or pencils on the desk. Our host may in fact have been a businessman, but this was definitely not where he spent his nine to five. No, this room—probably this building—had one use: secret meeting space. Alternatively, it was an ideal setup for getting the drop on an invited guest.

As Farid made small talk in Pashto, one of the official languages of Afghanistan, I rearranged my chair, turning slightly toward the double doors at my back, and the crackle in my brain eased just a little. I also found a spot in the window, just above the old man's head, where the doors were faintly visible in the reflection.

I ran through what I knew about the old man: He associated with members of Hezbi-Islami Gulbuddin (HIG), which sometimes functioned as a political party, sometimes as a terrorist militia newly aligned with the Taliban here in Nangarhar province, though I didn't know the exact nature of the man's involvement. He was—in some form or fashion—also profiting from narco-trafficking, though again, his role was unclear. He had met with coalition elements before and provided at least somewhat reliable information.

I didn't care about any of that. I'd heard a rumor that he'd once been a mentor to one of my senior contacts in the Afghan government—someone I was now secretly investigating due to his relationship with a brutal drug lord. That's why I was interested in this particular old man at this particular moment.

I wanted the least guarded answers I could get from him, so I first asked a variety of questions about all manner of things in Nangarhar. Then I invited Ellen to ask a few, which the old man answered. But he directed his responses to me, preferring not to acknowledge Ellen directly.

While he spoke, I reviewed in my head what to do if I heard or saw something indicating that those double doors behind me were opening. In situations like this, determining my course of action was like solving a complicated word problem. But here, with only one door allowing access to the building, it was pretty simple. I'd have my rifle ready for whatever came through that pair of doors, prepared to start firing and moving the three of us to the other door and down the stairs to Tommy and Ben, who were sitting with the Suburban, which was still running. The room wasn't large enough to sustain any exchange of fire—there was nothing to hide behind—and Tommy and Ben would never be able to get up the stairs in time to help us. And if we heard shots or explosions downstairs, same story. We'd be piling into that Suburban and getting the hell out of Dodge. In the first scenario, I'd send Ellen and Farid down first and try to cover them as they exited, and in the second, I'd be through the door first.

I thought about all of this while nodding and looking back and forth, from the man to Farid, as he translated a long, rambling story about the days of the Soviet occupation.

We hit a lull in the conversation, and I started asking more personal questions. I did this with a reverent, earnestly curious tone, and Farid communicated my meaning in a way that helped serve my goals. A good translator, like Farid or Salam, does so much more than convert language. They are ambassadors and cultural guides, and for an intelligence officer, someone who understands your aims can ease through barriers that you can't. Farid was clearly very good.

The old man—like nearly all humans—was charmed by flattery, and after a few minutes he was telling us all about his mentee.

It was clear that the Afghan official I was investigating back in Kabul was not a naive partner in his friendship with the drug lord.

That left two options: either he was personally invested in the drug lord's business or he was afraid of him. I suspected the answer was a little bit of both, but I wasn't yet ready to draw that conclusion.

After about an hour, the conversation became less fruitful, and both parties were ready to wrap things up. We exchanged the traditional salutations, shaking hands and then pressing our palms to our hearts, though I don't recall him doing so with Ellen.

The man walked us out of the building, and as we opened the door, Tommy shot me a look that said, *About damn time.* It had been such a valuable conversation that I'd gotten a bit greedy with the time. Because of this, our obviously coalition-owned Suburban had remained in one spot, signifying our location for quite long enough. "Let's pop smoke!" Tommy called out. We all climbed in fast.

But the old man lingered. He set his eyes on the plant and complimented it. When Farid translated the comment to me, I asked, "Should we give it to him as a gift?"

Farid sternly replied, "No, Jason, I don't think that's a good idea," and he didn't translate my question—clearly he'd just prevented me from committing a cultural faux pas.

A minute or two later, as we wove through traffic, I asked, "Farid, I don't get it. Why would offering the plant offend him?"

"It wouldn't," said Farid.

I was quiet while I waited for Farid to elaborate, but nothing came. Finally, after a long silence, I ventured another try.

"Okay, I'm still confused—"

"I like the plant," said Farid.

DIANA

When Jason told me about this day, I laughed about the plant and forgot the rest. Most of Jason's stories from the war were like this: entertaining and strange, but nothing really harrowing. He seemed to be taking it all in stride. He'd call and say, "You may have heard in the news about an explosion in Kabul, just wanted to let you know I'm safe." Or "Someone tried to blow themselves up close to our camp, but it's no big deal."

It was great that he was having such an easy time of it. Meanwhile back home I was miserable. The only way I could avoid the fear that he might be hurt or kidnapped was to pretend like it had already happened and try to go about my day as if he was already gone.

I never sent Jason a care package. His mom was so amazing about sending him something on a regular basis that I just let her take the lead.

I was twenty-five years old. I didn't have any friends who were married to the military, so I didn't have anyone to tell me to do anything differently.

Before Jason left, he told me stories of people getting killed overseas after a fight with a spouse over the phone. Unnerved, they'd been unable to stay focused and follow their training. They made a mistake, and that was it.

I resolved before he left that I would never do or say anything that might upset Jason when we talked. Even when we had normal relationship disagreements over the phone, I would just say, "You're right. Yes. I'm sorry." I didn't want to be responsible for anything happening to him, so I just let him win.

But the frustration lingered. I began a journal made up of nothing but my comebacks to his arguments, the words I held back. It was like I was in two parallel relationships—the one he thought we were having, and the real one, which existed only in my journal.

The military provided Jason with no mental health prep before Afghanistan and zero upon his return. Likewise there was nothing to help a reservist's spouse deal with deployment. Nothing to tell me what it might feel like, how to become emotionally prepared for it, or even what I would have to deal with upon Jason's return. In every other part of my life, I had been the planner who always stayed in control.

This deployment was the first time in my life that I felt powerless

and adrift, like I was up in some kind of runaway hot-air balloon I
did not know how to control. I just hoped I would land on the ground
in one piece.

———

About an hour later, we were on the outskirts of Jalalabad for mission number two of the day. I was standing at the top of a hill, behind a waist-high rock wall, looking down into a small valley crowded with a handful of modest homes. I could see two small children playing on one of the roofs. We were at the remote site the MTT had chosen to in-process and assess a couple dozen new Afghan National Army recruits. Tommy, Ben, and I were providing security on the perimeter.

The MTT was using a couple of little concrete buildings adjacent to a small mosque. The marines and Rambo had split the barefoot Afghan recruits (none appeared older than twenty-two or twenty-three) into two groups. The guys in the two-story building were giving their personal information to Farid, and those in the one-room, one-story building were finding out how many push-ups they could do in a row. Rambo had to teach a number of them what a push-up was.

Eventually, the senior local official—at least I assumed he was the head honcho—invited us to join him and his colleagues for lunch. A few of us walked with him to a rickety wood-frame building. We removed our shoes and sat along one side of a long table opposite our hosts. After the lieutenant colonel and the few low-level local politicians in attendance had exchanged unrealistically optimistic platitudes about the progress made by the Afghan National Army and Afghan National Police, I asked polite, open-ended questions about the border police, narco-trafficking, and the local HIG forces. But these local pols weren't about to speak candidly in front of one another. For me, group meetings like this got tiresome quickly; side conversations were impossible when only one translator was present. We ate nuts, rice, naan, yogurt, and goat. Well, I didn't eat the yogurt. I don't eat yogurt anyway, but yogurt in Afghanistan is made the old-fashioned way: leave milk out until it becomes yogurt.

Halfway through the meal, I went to the restroom. I squatted over a hole and tried not to let any part of me touch the walls, which were frescoed with brown brushstrokes. This was the very good reason I was carrying toilet paper in my cargo pocket and why no one at that table was eating with their left hand. My thighs burned with fatigue from squatting, but in my mind I laughed at the idea of having a business lunch with people who wiped their ass with a bare hand. Oh well. It wasn't like cowboys in the Wild West were carrying Charmin Ultra around with them either. Out here, even taking a dump was an adventure.

Once we'd returned to the top of our hill, we all admitted that we'd eaten very little. Even Farid, who was certainly used to the food, had been so busy translating, he hadn't had time to grab a bite. The recruits had lunched on rice and naan brought in by two old men in ANA uniforms, and they were still eating with Rambo when the marine first lieutenant pulled a few MREs out of his rucksack.

Rambo held court with the recruits, and we put the two ancient ANA soldiers, who had full bellies, on security duty. Then the rest of us ripped open our MREs, sat down against the side of a building, and had what the army mockingly calls an "MRE picnic"—trading for Skittles, peanut butter, and chicken tetrazzini, then stashing some of the cardboard containers in our cargo pockets to save for later.

Rambo and the marine gunnery sergeant had seated the recruits cross-legged on the ground in a shaded spot on the other side of the two-story building, where they wouldn't be seen (or shot at) from the hill across the valley. The recruits were completing what appeared to be a single-page test of basic literacy.

The ANA men came over and switched out with Tommy and Ben, and for a couple of minutes I listened to Farid translate their dirty jokes. Each time we laughed, they looked delighted. Then, as always happens with old Afghan soldiers, they started romanticizing the past, their mujahideen days of fighting the Soviets. As the con-

versation turned to what Nangarhar province was like before the
Taliban, the calorie-dense MREs and the gentle midday sun had me
fantasizing about a nap. I removed my fleece, and the warm air felt
good on my skin. It had been snowy and gray in Kabul lately, and
Jalalabad was winning me over.

Farid had gotten the plant out of the Suburban. He set it so it
faced the sun.

Reluctantly I stood up, ready to get back to patrolling the perimeter. Ben and the marine first lieutenant were walking my way and
joined me. Then I heard a distant crackling noise. I'd have known
it anywhere—the sound of a lot of rifles being fired. It was coming
from the other side of the hill across from ours, so whatever was
happening was about a quarter mile away.

At once the three of us ran toward the small rock wall that looked
down on the valley, diving to our stomachs and high-crawling on
our elbows and inner thighs for the last ten or fifteen feet, cradling
our rifles, muzzle off the ground.

For a few minutes we lay prone. We listened, barely breathing,
as the firing continued, interrupted occasionally by medium-sized
explosions that lightly shook the ground beneath us. Easily over
fifty guns were being fired. It is amazing what the adrenal brain can
accomplish in just a few minutes. Vast, graphically violent scenarios ripped through my mind, but not one of them involved running
away. I was straining to hear, trying to determine one thing: *Is it
coming closer? If it is, we're fucked. We're totally exposed, we're fish in
a fucking barrel, and we're grossly outnumbered.* And then I thought:
What if one of the recruits sold us out? What if they're here for us?
Objectively, I understood that I was supposed to be afraid. Dimly,
I *was* afraid. And yet some part of me, which had been growing
daily since the moment my plane touched down in Afghanistan,
was rooting for the fight to find us. My brain kept saying, *Okay,
this is it, let's fucking do this, let's go, let's GO.* I knew how insane this
feeling was, but I couldn't turn it off. It spread through my whole
body, radiating adrenaline. I no longer had a normal fight-or-flight
reflex. Nearly every day for the past few months, I'd been readying

my mind to kill, and by now the idea of option B had been trained out of me.

The firing had stopped. I summoned the nerve to crawl to an opening broken into the wall and peered through. I couldn't see anything. I turned to Ben and the first lieutenant and gave them a look that said, as casually as possible, *Hell if I know.* Who'd won? Were they moving our way? We waited a moment longer, and then the lieutenant colonel came walking right up to us, tall and proud, either unfazed or oblivious—I couldn't tell which. His first lieutenant said, "Sir, you can see fine from where Jason is; please get down."

"Oh, right, thanks," he said, and I rolled to the side to make room for him. We stayed in place but gradually got up onto our knees. Then all of us started talking among ourselves, speculating about what might have happened, settling on the theory that it was most likely an ANA unit in contact with HIG or Taliban fighters.

We were on MTT business, so the lieutenant colonel was in charge. He put everyone but Rambo and Ellen back on security and had the recruits hurry and finish the written test. The two dozen barefoot Afghan boys seemed understandably alarmed, but Rambo just collected the tests and declared them complete. Strangely, I caught myself thinking about the day I took the LSAT.

I knelt against the wall and kept an eye on the opening while snacking on a tube of peanut butter and some dry, chalky "bread," one of my favorite MRE snacks. Smoke was climbing into the air above the hill opposite us, and in the foreground I saw the little boys come back out on the roof and resume their game, as if a mere rain shower had passed through.

Having accomplished all we needed to at that location, it was time for the PSYOP portion of the day. Tommy had planned to put on an exhibition to recruit local high school students into the ANA—it would double as a show of strength for the benefit of the population. The MTT loaded all of the recruits into two pickups, and we proceeded into Jalalabad—a short convoy, with our Suburban in the trailing position.

The standard rule for convoys traveling through dense traffic is that you never let in any other vehicles. It's not like back home, where you just have to maintain sight of whomever you're following down the highway.

I'd been in convoys that crossed large, open areas with few other vehicles on the road. In those situations, you keep some distance from the vehicle in front of you, so that a single IED (improvised explosive device) can't take out two vehicles. But when traffic is dense, the threat is not bombs buried in the road. Instead, it's the possibility of being ambushed when stopped or slowed, or being blown up by a vehicle-borne IED. I'd commanded several convoys through traffic in Kabul, and a few times I'd even been behind the wheel. Staying right on the tail of the person in front of you, while you bobbed and weaved to avoid stopping, was fun—like a video game.

But in Kabul, when traffic came to a standstill, so did I; there was nothing I could do about it. You just sat there and waited, keeping your eyes peeled for any trouble. But here, in downtown Jalalabad, Tommy was determined to stay right on the tail of those recruit-filled pickups, and so when traffic stalled, he steered the Suburban onto the sidewalk, a move I'd not seen before. He also turned on a siren—I didn't even know we had one. Afghans were flying out of the way in every direction, while ahead of us, two truck beds of Afghan boys hung on for dear life, shrieking and laughing like they were on a roller coaster.

We drove out onto the soccer field behind a high school and sprang out as fast as we could to secure it. But it was enormous. The MTT told the recruits to run laps around the field while the rest of us, too few for our job, spread out as far as we could, forming an oval around the pitch.

An Afghan student, about sixteen years of age, came up to me. He looked me in the eye and—speaking in nearly unaccented English—addressed me as a peer, his equal: "I like America, my friend, but this is really stupid."

I invited him to elaborate, but he wasn't looking for my permission to continue. "You Americans are cowboys, and now my class-mates and I are in danger."

I didn't respond because I had no response. It was plainly obvious that our group was too spread out and too few in number to provide effective security. I tried to change the subject.

"I'm Jason. Do all of your classmates speak English as well as you do?"

He had no interest in offering his name, no interest in becoming friends. "Of course not. I study. Jason, you are being very stupid right now." He turned about, sweeping his hand to highlight how enormously exposed we were. "If you are always this stupid, you will definitely be killed by the Taliban."

I thanked him for the advice. He walked away, shrugging his shoulders as if to say, *Oh, well, I tried, but some people can't be helped.*

After about ten minutes, some of the Afghan recruits began to tire. No one had told them, say, to drink water before they ran. Their performance didn't add up to much of a show of force, and I doubt anyone watching felt reassured that the Afghan military had a bright future. Few of the recruits had shoes, and all were skinny; their faces looked both childlike and old beyond their years. We didn't know why they had joined up. Maybe it was patriotism, maybe money. We didn't care. Training these kids was America's ticket to getting the hell out of this war after half a decade of fighting. But as I watched them, it didn't feel like that ticket would get punched anytime soon.

Tommy apparently felt he had created enough of a spectacle to get people talking, so he yelled for us to mount up. Each of us echoed the command so that it could be heard across the pitch. Less than a minute later, our Afghan circus troupe hit the road again.

For a couple of hours we accompanied the MTT as they drove the recruits back to their pickup locations. Finally, the workday was complete, we were hungry again, and Ben mentioned he'd heard good things about the food at Jalalabad Air Field. After a goat-and-MRE lunch, dinner at JAF sounded much better than whatever we'd get back at the FOB. But the sun was setting, and it wasn't advisable to travel at night, especially in a single vehicle and without night vision goggles. The lieutenant colonel opted out, and his team headed "home" to the FOB.

Tommy's mind was set on dinner at JAF. "I have the number for the ODA team here," he said, as he patted his cheap plastic cell phone. "If anything goes down, I'll just call them." ODA stood for Operational Detachment Alpha, a twelve-man special forces team. And so the six of us—counting the plant, which by now pretty much outranked me—set out to find JAF.

A half hour later, it was pitch-dark. We bumped along on deserted dirt paths bereft of street signs, and the vibe in the Suburban had gone from *Black Hawk Down* to *Harold and Kumar Go to White Castle*. After the tenth time Ben looked up from the map and said, "It's got to be right around here, right?" we realized we were lost. And vulnerable.

"I swear to God," I said, "if we die on account of wanting slightly better shitty food . . ."

Everyone in the car burst out laughing. Finally, Tommy swallowed his pride and called special forces—we needed directions.

We finally got there. JAF's chow hall was just like its cavernous counterparts stateside: big and crowded, with fluorescent lights and linoleum floors. Row upon row of uniformed soldiers were eating dinner. As we shuffled through sideways with our trays, unshaven and out of uniform among the many well-kept high and tights, I could feel all the eyes follow us as if we were desperadoes riding slowly through a frontier town. Everyone looked at us like we were the most Hooah motherfuckers in RC-East. I knew they thought we were "operators," and I strutted around like a returning all-star on the first day of spring training.

And yet, even though we were in a chow hall on a heavily fortified American installation, I spent the whole meal watching hands, tracking door locations, and keeping an eye on anyone remotely out of place. I couldn't help it; if anything, I was reassured by how thoroughly this practice had become second nature.

Other than praising the food—we agreed it had been worth the risk—we ate in silence because we were hungry and because we'd been together since dawn. I was thinking about what we'd done all day, how badass it had felt and yet how unremarkable it turned out to be. Badass because, as on every day of my deployment so far, I'd seen and done things my friends back home would find impossible. Unremarkable because I had done some of it before, but also because when I really thought about it, nothing had actually *happened*. I hadn't been blown up and I hadn't fired my weapon.

I do like the weather here, though, I thought. *On my next deployment, I should try to get assigned to RC-East or RC-South, for the weather alone.*

When we got back to our little FOB, we each took our five minutes on the Defense Switch Network phone. Usually I called home after breakfast, which meant I was waking Diana up shortly after she'd gone to bed, but at this hour she was fully conscious when she picked up. As usual, I couldn't say much about my day, but I wanted to tell her, eventually, all the stories I could. So I'd often say, "Remember to ask me about" this or that "when I get home." I assumed she was bookmarking these reminders and eagerly awaiting the reveals. There was probably a list. She liked lists. We'd go over all of it together, and she'd understand. I hated the fact that I couldn't tell her about my days as they happened, but she knew that this secrecy was important.

Mostly I just wanted to hear her voice. When I got some gossip about family or friends, it was morale-boosting, a bonus, because it kept me from feeling quite so far away. More than anything, I was proud of how well every phone call home had gone. Not once had Diana and I disagreed about something.

We'd been thousands of miles apart during college. When we were under the same roof in law school, we'd fantasize about our future. This was not talk about babies or houses or cars or vacations. We spoke about campaigns we'd run and offices we'd win and laws we'd change. But over the past year, we'd barely spent two months together. I was eager to make up for lost time.

After we'd made our calls, everybody was ready to rack out, but I was wired. "I heard a rumor about a PlayStation in the MWR hut," said Ben. He was referring to the morale, welfare, and recreation hut. I grinned. I had been jonesing for that ever since my mom shipped a PlayStation to Camp Eggers and I'd immediately fried it, forgetting to use a power adapter before I plugged it into an outlet in my safe house.

The one-room hut had a floor covered in dirty rugs, a couch, two beanbag chairs, a couple of bookshelves full of worn paperbacks that had never made it home from their deployments, a television, and, lo and behold, a fully functioning PlayStation. Lounging in the beanbags, we played *Madden* football, rifles on our laps, and talked about home and how great the weather was on this side of the J-Bad Pass. *I am a fucking cowboy*, I thought. I couldn't believe I had to go home in a few weeks.

3

RUNNING ANGRY

When I'd deployed to Afghanistan, I'd flown in on a C-130 from Doha, Qatar. I was smooshed in so closely with dozens of other soldiers, it made flying in the middle seat of a commercial flight look luxurious. The day I left, four months later, in early February 2007, there were maybe ten other people on the plane with me. I spread out my gear on the floor, lay down, and fell fast asleep.

The minute we landed at Al Udeid Air Base in Doha for decompressing and debriefing, I noticed a weird little twitch, like a muscle spasm, in my left eyelid. *Huh*, I thought, *that's annoying.* Two days later, when I got home to Kansas City, it was still there, but I figured it would go away after a few days of genuine rest.

Six months later, it hadn't left me, and later, when something stressful happened, like if I was in a crowded, noisy place or writing on a deadline at my job as a lawyer, my right eyelid would get jumpy too, spasming away at the edge of my vision.

When I went to talk to a doctor about it, he said, "I suppose you bottled up all that stress for four months and now it's just sort of working its way out a little at a time."

I wanted to say, *No shit*, but instead I thanked him and handed over the co-pay.

Nonetheless, after I came home to Kansas City, I channeled all my energy into a new and exciting mission. Back in 2005, I'd registered a committee to raise money for a campaign for the Missouri state legislature, and I'd also founded my own PAC. When I got home, Diana was managing a campaign for city council, but once that ended, in April 2007, we dove into the state rep race.

Since my district was heavily Democratic, the real campaign was the primary, not the general, which meant I had to move fast. That deadly serious fifteen-year-old was still in there, champing at the bit, but now he was a twenty-six-year-old soldier who embraced suffering as a virtue. The army had taught me that any skill set could be gained with enough training, so I decided to take on the political equivalent of "warrior tasks and drills." Battle drills and troop-leading procedures were replaced by call time and canvassing.

Sure, I'd had some adjustment issues when I finally came home from Afghanistan and transferred from the Army Reserves to the Missouri Army National Guard. But after a few months, my heart stopped racing every time I got into the car, and Diana and I figured the problem was behind us. I was having some nightmares, but we chalked those up to stress and figured they'd fade with time.

Most important, after close to a year and a half of active duty in Arizona, Florida, and Afghanistan, Diana and I were fully reunited. Where professional life was concerned, we were still just a couple of kids, but we'd been together for nine years, and throughout that time, we'd shared two goals: be together and try to change the world. We were now doing both, and it was an absolute love affair.

Shortly before I came home, Diana left her law firm. She'd been a rock-star first-year associate, exceeding her billable hour requirements *and* bringing in clients (something first-year associates just don't do), and she'd been offered a big pay bump to stay on. But she'd always had her eye on being an entrepreneur.

Diana had grown up watching her parents claw their way out of poverty by starting a small business, and she was brimming with ideas. So just as I was launching my first campaign, she was launching a new business. She was in her element, hustling and selling, learning at an unrelenting pace. We called ourselves Team Kander, and Team Kander was having a hell of a good time kicking so much ass.

As hard as I'd learned to work in Afghanistan, Diana matched me step for step. At night, we'd stay up, planning the campaign and playing with our two energetic dogs, Winston and Shelby, in a house we'd barely furnished (and never would, really). All we did was work and make love.

During the day, I was a corporate lawyer, required to bank as many billable hours as I could, but I was only mildly more attentive to that work than I had been in law school. One weekend a month and a few weeks a year, I was a platoon trainer for the Officer Candidate School at Fort Leonard Wood, and when I wasn't campaigning, I was planning the next drill weekend. The army remained the best part of my professional life. On morning runs during those weekends, the platoon would be singing cadences, and between verses I'd be yelling at them about how great it was that at the end of this, we were all gonna get to eat *Army Breakfast*.

Diana's day job—getting a business off the ground—had much higher stakes than the question of how many billable hours I'd racked up, but she treated my campaign like a second full-time job. We wrote copy for the website, even though it hadn't launched. We made endless lists of people who could give money or time, and we created our budget. Diana even taught herself Photoshop so that, in my campaign materials, it would look as though I owned a full professional wardrobe and not just one ill-fitting brown suit. But all the lists and Photoshopped mailers in the world weren't going to solve one big problem: I was twenty-six years old and no one knew who the hell I was.

Campaigns at any level present an almost beautifully simple problem: how do you contact and convert as many voters as possible to your cause? In large, well-funded campaigns, the machinery dedicated to solving this problem is as vast and complex as any military operation, using any means possible—TV, radio, billboards, digital media, yard signs, cameos in movies, roaming vans equipped with PA systems, huge cardboard cutouts, video game avatars, anything we can think of—to reach voters' eyes and ears. (We'd target the other senses if we could—imagine peel-off fragrance samples, like those used to promote perfumes, to pitch a candidate's pleasant personal scent.)

But the farther down the ballot you are, the fewer tools you have. At first, I didn't have the money for much of anything—I didn't even have staff until a few months before the primary, and by staff I mean we hired one fantastic young person, Margaret Hansbrough, and paid her not very fantastic wages. But I did have one thing the people I was running against didn't have: really comfy standard-issue army boots.

The best way to solve a simple problem is with a simple solution. To win, I had to run in a way that no one else would, or could, compete with. I knew three things: I didn't get tired, I didn't mind suffering, and there were about eight thousand primary voters in my district. All I had to do was meet every one of them.

On August 1, 2007, I laced up my boots and started knocking on doors. It turned out that walking for miles every day was easy when I wasn't shlepping forty pounds of gear, and meeting new people was more comfortable when I wasn't worried about getting beheaded. Diana came along to knock on the first several thousand doors with me, but once winter came, she would stay in the car, following behind me in her motorized command center. When I finished a block, I'd hand her a page of voter data for the spreadsheet she'd designed, and she'd enter it into her laptop as I started on the next block. We tried to be stealthy about these door-to-door efforts. We never posted about them online. By the time my opponents knew I was a serious contender, we'd already hit every house in the district at least once.

At one door in Kansas City's Waldo neighborhood, I mentioned I'd been an army intelligence officer in Afghanistan, and a woman said, "I wonder if you met my daughter's brother-in-law, he's in army intelligence too." People did this all the time, and I'd politely ask for the name, pretending the army wasn't enormous. But to my surprise, she mentioned my friend Kevin, from the tactical human intelligence team.

What were the odds that I'd actually know this person, and that it would be somebody I'd worked with closely? Other than Salam, I hadn't stayed in touch with anyone from the deployment. I'm not proud to admit it, but I hadn't stayed in touch with anyone who wasn't a prospect for a campaign contribution. This meant that my classmates at Georgetown Law were hearing from me a lot more than my classmates at the Hoya Battalion, though many of the latter were still overseas and needed to hear from friends back home.

This pleasant woman told me Kevin was currently stationed at Fort Leavenworth, which was only about forty-five minutes away. I left with Kevin's number and a promise to call him. "I hope you will," she said. "I think he'd want to see you."

Months later, I made a second pass through that neighborhood, and when I came to this same door, I realized I hadn't reached out to Kevin. So I began the conversation by apologizing and asking if he was still at Leavenworth. "Oh, honey, I'm so sorry," she said. "Kevin was killed in a car accident a few weeks ago."

I was stunned. Apparently, he had been alone in the car, and no other vehicles were involved. But I'd driven with Kevin, knew how careful a driver he was—he was the guy actually using his blinker in Afghanistan. How could he survive being on the road there every day, just to come home and die in a single-vehicle accident? It was so random and cruel. *It just doesn't seem fair*, I thought. And it didn't make sense.

All of a sudden, the whole business of knocking on doors—my whole campaign—felt trivial and selfish. It had begun to rain. But all I could do was step down from the porch, walk to the next house, and knock on the door. It wasn't just that I was determined to win. I needed a distraction from the anger I felt toward myself for not calling my friend.

I hadn't seen Kevin in over a year, but I remembered that he was married, just like I was. I thought about his wife's tremendous loss—trying not to picture Diana going through the same ordeal. But I fought off those feelings as hard as I could. I didn't feel I had earned the right to mourn Kevin's death, I had not been enough of a friend, so I refused to indulge in sadness. Instead, I did the only thing

I knew could suppress those complicated feelings: Work harder. Do more.

I want to stop here for a second. I feel like in the past, when I've described my work for this first campaign, it winds up sounding easy, as if I didn't feel the strain or the pressure or the ache. But I did. Constantly. The campaign consumed me. I knew how long a shot victory was—I had two opponents with far better institutional support. But I also knew I had to win. I was unable to imagine what life would look like otherwise, and I didn't want to imagine it. To some degree, all candidates go through this, but for me, it went deep. I would go to bed at night dreading the prospect of failure in a way I had never fallen asleep dreading anything—not even the enemy in Afghanistan. The dread was yet another reason why sleep didn't come easily. It wasn't just a fear of losing but also a new sense that everything I engaged in was all or nothing.

The campaign lasted a year and six days, and the whole time, I ran *angry*. I worked every single day, and by work I mean I hit the pavement and shook hands and had doors slammed in my face and let bugs feast on me. I barely slept, in large part because of a new and unwelcome part of my life: nightmares. The Taliban or Al Qaeda would crash in and throw a hood over my head and wrench my arms behind my back. I'd be out on a convoy and it would blow up. Basically everything I'd feared during daylight in Afghanistan now happened to me at night in Kansas City. I'd never had nightmares before or during my deployment, but now I had them every night. I developed sleep paralysis, so even when I did wake up from a nightmare, I still had the sensation that something huge and terrifying was looming in the bedroom, and I'd be frozen in place, helpless. I'd try desperately to wake Diana by grunting and gasping, so that she could shake my body until the paralysis ended. And then I'd get out of bed and stalk the house, armed with a Louisville Slugger, checking every door and window.

Other tics had started too. When I went out to eat with family or friends, I made sure I sat facing the door. This was absolute:

if I couldn't see the door, a low drumming would start on in my brain, a sensation like the vague dread you get when you know you have ten different tasks you have to do, but you can't remember one of them—and if you don't do that one task, you could die. My first Fourth of July after Afghanistan was loud and uncomfortable, and driving was sometimes a challenge: I'd find myself swerving to avoid things. I told myself this all made sense—anyone who'd spent months on high alert, expecting an explosion at any turn, would feel twitchy behind the wheel for a while.

Honestly, I was ashamed of myself for all of this. I knew how common these experiences were among combat vets, but I was—to my thinking—absolutely not a combat vet. Still, I did at least look up the symptoms of post-traumatic stress, if only to convince myself I didn't fit the profile. A four-month tour wasn't enough to mess with a person's brain, I believed, especially not when that person had the advantages and support that I did. To me, nothing I experienced counted as "trauma." For one thing, I had never been in a firefight. All I'd really done was go to meetings, and now it was frustrating and a little embarrassing that every time a voter invited me into their home, I found myself tugged back to some meeting in Afghanistan. Often, if it was cold outside and I was seated on a living room couch, I'd be thinking about a man named Sabet.

Abdul Jabar Sabet was a tall, richly bearded Pashtun who, in 2006, had recently become the attorney general of Afghanistan. We referred to him simply by his last name. He'd become my favorite contact in the Afghan government: his English was great (he'd worked for Voice of America, the international broadcaster, in the States before returning home), he was influential, and as far as I could tell, he had no reason to help anyone kidnap me. Unlike most Afghan officials, Sabet appeared to be a paragon of morality: he'd even spearheaded an anticorruption campaign.

Other intel people had warned me not to trust him too much, given his ties to the leader of the terrorist group HIG. But I wanted to believe in him—and if there was a chance that his anticorruption efforts were real, I wanted to support them.

He seemed to like me too, and despite the fact that I was just a low-ranking kid, he never turned down my requests for a meeting. Eventually, folks from other agencies got wind of our relationship and began to ride shotgun with me when I went to pay a visit.

On one particular day during my tour, I'd shown up with two guys from the Defense Intelligence Agency (DIA), with their translator in tow, and right from the jump I knew that something was off. Approaching Sabet's little building in Kabul, we drove through a rickety gate and were greeted by six men in Afghan Border Police uniforms, their AK-47s at the low-ready position. I'd never seen border police here—Sabet had his own security. They started barking at us even before we got out of our little Mitsubishi Pajero.

I realized I was holding my breath as the DIA translator said, "They want us to leave our weapons and the body armor in the car." In normal life, when something unexpected happens—say you get home and realize your TV was never turned off, or there's a strange car in your neighbor's driveway—you process it, shrug, and move on. But in war, when something unexpected happens, you fixate on it. We're taught to "expect the unexpected," but what that actually means is "when the unexpected happens, prepare to fight like hell."

It was normal to be invited to remove body armor—it was rude to visit someone dressed like you expected him to blow you up. But leaving my weapon behind? The DIA guys might have thought I was so green that I'd comply, but I knew they wouldn't dream of it, so I took my 9mm Beretta out of the body armor holster and—instead of transferring it to my hip holster—tucked it into my waistband, under my sweater.

Inside, the border police walked us past the staircase I'd normally ascend to meet Sabet in his office. They led us to a long, narrow room with a big horseshoe couch at the far end. They told us to sit. On the other side of the room, Sabet was holding court—literally. For an hour, we watched as a shuffling parade of Afghans from all over the country came before him to present their complaints. I sat there, writing it all down—the man who raged against the corruption in Mazar-i-Sharif, a woman whose new boss forbade her to come to the office in a burqa—concerned that our wait would be so long, we wouldn't get to talk with Sabet and the DIA guys would be mad that I'd wasted their time.

But when the litany of complaints at last ended, Sabet came over to us and chatted awhile, explaining the cases he'd just handled. Then he said, "I have an old friend visiting today, and he has asked to meet you." Three of the border police entered the room and stood facing us, still toting their AKs, followed by a man in a border police uniform so snappy and crisp, you could have bounced a quarter off it. His beard was sharply trimmed and his teeth were in excellent condition, a rarity for Afghan men. I knew I'd seen this man before somewhere. He greeted Sabet without looking at us. They spoke for a moment as we sat there, trying to figure out who this was—he was way too neat and clean to be an average member of the Afghan military.

Eventually Sabet turned to me. "Jason, this is my dear friend Hajji Zahir."

I tried to control my facial expression. I shook Zahir's hand, hoping he didn't notice that mine had gone cold and shaky. Then we made the Afghan gesture of greeting, placing our hands on our hearts. I caught the DIA guys' eyes, and they were all asking the same question—*What the fuck is happening?*

We all knew General Hajji Abdul Zahir Qadir. His father had fought against the Taliban, and they'd assassinated him in 2002. Zahir had risen to his post in the border police thanks to extensive patronage, but lately he was moonlighting with the terrorist arm of HIG, which had not only sworn allegiance to Al Qaeda but was now in bed with the Taliban too. We didn't think he himself was Taliban—they'd killed his father, after all. Instead it looked like he was making tons of money running drugs along the border. We were actively investigating him and considered him a high-value individual—meaning he was cruising toward a possible arrest, or worse.

Now I was having tea with him. And his men had guns much bigger than ours.

Steam curled from our untouched teacups as Sabet and Zahir talked about how they knew each other—they were from the same province. Zahir ate the nuts and dried berries that Sabet's aides had laid out. Sabet, who was usually a big talker, had gone quiet. This was Zahir's meeting now, which wasn't a coincidence. This encounter had been set up so that he could be in the room with us.

My heart was beating out of my chest.

He knows what we do. He knows we're investigating him. Maybe there's something he wants to hear, and if he doesn't hear it, we're going to be taken away and killed?

Zahir spoke mostly to the translator, not to us—he seemed to be keeping a close eye on the man to make sure he was translating correctly. As I sat there, plastering the occasional smile on my face, my brain was running at maximum warp, trying to figure out what the hell Zahir was doing here with Sabet. Was Sabet naive about his corruption? Or was he involved with it?

Zahir was beginning to get worked up. "The Americans, the foreigners," he said with sudden fury, "they don't *understand*." He was yelling at us. It wasn't simple venting, like when you wind up yelling at a representative of an insurance company—the person has no real power, but since they're the one at hand, you shout at them. No, this man wasn't just grousing. He knew we had power—we were the ones investigating him. He sounded like someone who'd come prepared to kill. I felt real fear.

Another possibility arose in my mind: maybe Sabet wasn't the man I'd believed he was. And if he wasn't, and this was a trap, there was only one way to get out alive: be the first one to shoot. I couldn't afford to be wrong—if I went for my pistol, there would be no going back.

I began making mental calculations. If Zahir turned and said one word to his goons, I would pull my weapon and fire. But we were seated. They were standing. There were four of them, and four of us—but at least three more of Zahir's people were in the hallway.

Zahir's men in the room with us had their hands on their weapons. Ours were tucked away. I was sure the DIA guys were doing the same math I was. I started working out what to do if I saw one of my guys go for his concealed weapon.

Momentarily, I asked myself, *Am I even* allowed *to shoot these guys?* And then the thought was gone—it wasn't going to matter. I imagined the feel of the Beretta in my hand as it fired, tried to visualize everything—two in the chest, one in the head. I picked out the first goon I would take out, the one directly across from me, knowing the guys next to me were thinking the same thing.

Then, after about forty-five minutes of deliberately scaring the hell out of us, Zahir got to the point. He started to spill the beans on every warlord "the Americans must arrest to end the heroin trade." I could have laughed. Suddenly I understood why he was here. Intimidate us? Sure. But get rid of us? No, he needed us to get rid of the competition. I had to admire the chutzpah, but we already knew the information he was dishing out. Still, we very obsequiously wrote down every name he volunteered, bathed in relief that we were going to walk out of the room safely.

Outside it was cold, the ground caked in sleet and frozen mud. The DIA guys and I walked carefully back to our vehicle. I put my body armor back on and reholstered my pistol. Then I saw the DIA guys pick up their body armor too—with their weapons still attached. As it turned out, they hadn't brought them along.

I'd been the only one of our team in there who was armed.

A wave of nausea crashed over me, almost making my knees buckle. I had been so close to pulling my gun out, and if I had, every one of us would have died.

Sure, I had some scary moments like that one over there, but the worst had never actually *happened.* I'd concluded that I merely had something called "battlemind," and already it was starting to go away. The eye twitch eventually faded; certainly the other issues would too. Sure, sometimes I had trouble controlling my temper. When a conversation with Diana unexpectedly veered into a fight (which had been happening more and more), she'd tell me to go downstairs and take my anger out on the heavy bag, which she had installed for letting off steam while I was away. But my behavior made sense. I'd been to war. And more than that, I'd kind of *loved* the war.

After a while I figured the problem wasn't "battlemind" alone—it was also some type of survivor's guilt. As happy as I was to be back with my family, it had been hard to leave Kabul. The day before I got on that C-130 at Bagram, I ran into a guy I'd worked with at the JIOC, a navy chief petty officer who was returning from his midtour leave. He had done six months in Afghanistan and been home for two weeks; now he was getting ready to convoy back down to Kabul for another six months. Like a lot of my friends, he had been there when I showed up, and he was still going to be there for several months after I left.

I confessed that I felt pretty lousy for leaving after only four months. In fact, I was ashamed of it. I knew how much I had to look forward to, and I had been on active duty away from home for most of the past fourteen months, but the truth was, a four-month deployment was pretty small potatoes. "I should have done so much more," I said. "It feels wrong to leave."

"Look, man," he said. "Everybody knows somebody who did more. Go home, sir."

I'd had a similar conversation with Colonel McCracken earlier that same week, and he'd said it was time to get home to Diana. Both conversations made me feel better momentarily, but only momentarily. What they hadn't said was "You've done enough." In fact, I knew I hadn't. Every time I got into my little Mitsubishi and drove through those gates, I knew I could always do more because I *had* to do more, because that was the only way I could survive. In Afghanistan, it meant being more vigilant, more prepared, more equipped to get out of some meeting alive if things went south. Now that I was home, it meant doing more to win, more to prove to myself that I was more than just a guy who'd gone to meetings for four months.

The truth is, I was mad as hell—like, genuinely angry—all the time. And being angry was strangely comforting. When the rage bubbled up, this unearthly calm would settle over me. Before I'd even announced I was running for office, I was out at a bar one night, and someone said, "At least you weren't in Iraq. Is there even anything going on in Afghanistan anymore?" Suddenly I felt viscerally enraged—more so than the thoughtless comment warranted. I wanted to tear the guy's arms off. This was the exact thing everyone deployed in Afghanistan feared Americans were thinking—and to hear it just casually tossed out there . . . suddenly I was seized by the kind of fury I'd felt in Afghanistan when people needlessly put others at risk. But I kept it together; all I did was growl that the per capita casualty rate in both wars was still the same. And then, being me, I wrote an op-ed in the *Kansas City Star.*

That fury rippled out to everything and everyone who appeared to be in my way. I didn't just want to beat the people running against me in the primary, I was furious that they were going up against me at all. I knew I was young—I'd gotten hardly any endorsements, the lifeblood of the unknown, down-ballot candidate. To me that meant I had to prove my opponents' endorsers wrong. Every night I'd go to bed asking myself, *Did I do enough today to help myself win?*

One day, late in the campaign, I was knocking on doors and feeling truly out of gas. The summer sun was baking the pavement. I had lost the weight I'd gained back—and needed to gain back—after I'd gotten home. I weighed 140 pounds—I looked unwell. That day I was in a neighborhood called Santa Fe Hills, which isn't just a glossy real estate euphemism—the hills are plentiful and steep. I was standing at the bottom of one, looking up, squinting against the sunlight. This was my third pass through this working-class neighborhood, and I was trying to make sure I'd gotten to every single voter I could. The thing was, I'd just found out that I was actually winning this thing. A friend at a polling firm had gotten me a deep discount for a poll, and he'd called me just an hour earlier with the results, which showed that I was up by a lot. I could have called it a day. I'd already talked to almost all of these voters—twice! What was one more trip going to do, other than dehydrate me? Why do more? This wasn't Kabul.

But stopping never seriously crossed my mind. I had become a soldier. I might have worn the uniform of a politician and lawyer during the day, but I still saw a soldier in the mirror. It was the thing I'd worked at the hardest, that I'd sacrificed the most for. I was part of the US Army the way I was from Kansas City, or married to Diana: it was who I was. And if I wasn't a soldier, I wasn't anything.

Could I have gone to bed that night and reassured myself that I'd done enough even if I'd wimped out on that hill, the kind of steep incline I made my officer candidates sprint up in PT? That hill meant only one thing to me: a chance to run up the score. I didn't just want to win. I wanted to make my opponents—who were perfectly lovely people, by the way—pay for even thinking they could beat me.

On election night—August 5, 2008—my family and friends (including dozens of people who had answered when I knocked at their door and now were dedicated volunteers for Team Kander) came together at a little bar in Waldo to watch the returns. As it became clear that I was going to win, I got up on a chair to give a speech, thanking everyone. You see politicians, regardless of their party, do this a lot on election night. I can promise you that these are among the most honest, heartfelt speeches we give. And it's not just because the votes are in and there's no one left to convince. At the end of a grinding campaign, you realize just how many people had put themselves on the line to work for you and dedicated themselves to the insane proposition that *you* should be vested with power. And this wave of gratitude just crashes over you, as if it has mass and speed; it almost takes you off your feet.

When I got up there to speak, I saw a room full of exuberant people clad in T-shirts bearing my name. I can't say I was humbled. I was too proud of myself to be humbled by it. But somehow, at the same time, I felt this pang of unworthiness. Looking out at my family, at Diana, my mom and dad, my brothers, my grandfather, everyone who had all pulled hard to help me win, I became so choked up, I could hardly get the words out. But it was more than the emotion of the moment. I was exhausted. Try to remember the last time you stayed up all night—how fragile you felt around dawn, how vulnerable, like if someone so much as touched you, you'd burst into shards. Election night was like that, only I'd stayed up for a year and a half. I was one big frayed nerve.

Throughout the campaign, I'd put faith in the idea that winning would not only make me happy but would also alleviate my "little" problems, such as anger and insomnia. "It'll be better when I win," I'd say to myself about damn near everything. And after I won, it seemed, for a few days, that I'd been right.

And then—as would become my practice—I commenced moving the goalposts. Now that I had won, it was time to be the savior, the one to fix all of Missouri's problems. I believed I had to achieve something transcendent and meaningful with the same urgency that I'd believed I'd had to win. And just like that, the anger and insomnia returned. I had put myself behind before I'd even begun.

4

THE SHAME SPIRAL STAIRCASE

From the moment I was sworn into the Missouri House of Representatives, I was on a full-blown crusade. Diana was 150 miles away when the legislature was in session, so I felt no incentive to "come home" to my empty Jefferson City apartment. I was in my office day and night. I'd vowed that I would personally answer every single email from a constituent. I felt I owed it to them. Besides, the alternative was sleeping, and I knew what was waiting for me if I nodded off. Diana was too far away to rescue me from sleep paralysis.

Not one of the other 162 people in the Missouri House of Representatives was working as hard as I was—proof to me that they weren't really serious. I saw my compulsive behavior as noble and good, saw it as the fuel that got me elected. Meanwhile, instead of abating, the anger that I had experienced throughout my campaign burned even hotter as I discovered how futile it can feel to be a representative in the lower house who belonged to the minority party.

The legislature was in session for five months of the year, and in 2009, when I was sworn in, the annual salary was $35,915, with a decent per diem. Rather than take some enjoyment in my new role, I felt hypercritical of those around me. To me, the chamber was full of dilettantes for whom the worst thing that could happen, in their entire lives, was losing an election—and I'm just referring to most of the members of *my* party. The other side—from my perspective as a twenty-seven-year-old flaming sword of legislative righteousness—was a bunch of flag-pin-wearing chicken hawks who treated corporations like people and Black people like three-fifths of a corporation. And I knew how they saw me. I was this hotheaded kid from a solid-blue district who needed to learn a thing or two about how things worked in Jefferson City. Luckily for me, I wasn't the only such kid.

Stephen Webber was elected alongside me in 2008. Two years younger than me, he was a veteran too, and—to my thinking—one with a whole lot more cred than I had. Stephen was a marine re-servist from Columbia, Missouri, who'd done two tours in Iraq and seen extensive combat; he'd guarded the perimeter (not the inside) of Abu Ghraib in 2004 and led an infantry squad in Fallujah in 2006.

He reminded me of my friend T.J. Hromisin.* T.J. had slept in the bunk next to me at Fort Lewis. A dynamic leader with a sense of humor and an electric smile that lit up his whole face, he wasn't just the best in our fire team or our squad; he was the most capa-ble cadet I'd ever seen. T.J. led our platoon's first training mission and set an impossible standard. The captain evaluating us called his performance one of the best he'd ever seen by a leader at any level. "Cadet Hromisin," he said, "if I were given an infantry company in Iraq or Afghanistan tomorrow, I'd ask the army to give you a direct commission and assign you to me, so I could put you in charge of one of my platoons immediately."

A year later, an Iraqi sniper shot T.J. He survived, but he was now blind, brain damaged, medically retired, and nothing like the funny, fiery kid who knew every word of "Baby Got Back." I thought about T.J. a lot, and being around Stephen was like being around the T.J. I'd known at Fort Lewis.

If you asked, say, a hundred vets to list the ten best things about being in the military, I'm pretty sure that every last one of them would include the way in which being in the service bonds you in-stantly to total strangers. The moment you realize that you both speak the same language, that you've both been to war—doesn't

---

* Pronounced "Ro-Mission."

matter which one—it's like being in a distant country, like Bhutan, and running into a person who grew up right down the street from you. In Stephen's case, we didn't just bond, we practically merged. Like me, he'd also been dismissed as a candidate because of his youth and went on to beat the snot out of his opponent in the primaries. And he was every bit as angry as I was.

We'd sit together in Democratic caucus meetings and listen as people got up and said things like "we're all in the trenches together" or "we need to hold the line." Over and over they'd unthinkingly spout these clichés borrowed from war. It was ridiculous and—to our ears—out of touch. Every day, among our colleagues, things we found trivial were spoken of as life-and-death matters. We did our best to laugh it off, but we weren't always successful. Gradually, it became more difficult to tolerate.

Over time, Stephen and I became so pissed that, inevitably, the day came when we let it rip.

In the early days of the Obama administration, one of the biggest controversies was Gitmo—the prison at Guantánamo Bay. President Obama was trying to close it (he never succeeded), and Republicans had decided to turn it into a wedge issue all the way down to state politics. At the statehouse the GOP decided to take a Bold Stand by passing a nonbinding resolution to bar any Gitmo prisoners from being transported through or over Missouri on their way to the military prison at Leavenworth. The bill was purely for show: states don't get to dictate who gets housed in federal prisons or where the federal government's planes can fly.

This bunch of Missouri politicians acted like they were taking it to Al Qaeda by showing our state to be so scared of detainees in shackles that we wouldn't let them be flown overhead at thirty thousand feet. The hypocritical speeches in favor of "supporting our brave troops" and "standing up for America" really set off Stephen and me.

No one dared oppose the move. So Stephen and I got up, one after the other, and tore into the people around us. We asked sarcastic questions ("Once the terrorists hear about this resolution, how soon will they surrender?"), mockingly flipped the narrative ("Why should Missouri be softer on Al Qaeda than Kansas is?"), and made some direct accusations ("It takes a real coward to support a war and screw over the people fighting it").

We had just come from jobs where everyone was willing to give their life for one another, and now we were surrounded by these dishonorable and self-interested hacks. To our way of thinking, every cynical piece of legislation aimed at helping the powerful at the expense of the powerless was a slap in the face of our friends who were still fighting for America. We felt that no one knew what we knew because they hadn't seen what we had seen; every person who opposed us became an avatar for the people who had pushed a war of choice in Iraq but denied our troops in Iraq and Afghanistan the body armor and equipment that they needed. All Republicans were Donald Rumsfeld until proven otherwise.

One night, about a week after Stephen and I took office, we left the capitol at about 1 a.m. and went to the McDonald's drive-through. (There's not a whole lot open in Jefferson City at 1 a.m.) We ordered our food, and then Stephen said to the guy in the window, "You know what? I'll take one of those sundaes."

I looked at him, skeptical. "Wow. Ice cream at one in the morning?"

"Listen, man," he said. "Ever since I got home, I've just decided: I'm gonna be comfortable and I'm gonna do what I want. If it's cold, I'm gonna wear a heavy coat. If it's one in the morning and I want some ice cream, I'm gonna have some fucking ice cream."

That was what separated Stephen and me from the rest of our colleagues: everything we did was something we did "since we got home."

In the legislature I found my calling: ethics reform. It earned me the loathing of pretty much the entire assembly. Leadership refused to recognize me on the floor, killed any bill I filed just to watch it die, and cooked up bullshit ethics complaints against me. Even people who did support my ethics bill told me they couldn't do so publicly for fear of losing a committee assignment or a nice office, or even a parking spot. A fucking *parking spot*. I'd seen nineteen-year-olds, ready to puke from fear, climb into unarmored SUVs to travel roads that might literally explode, and now my colleagues were worried about having to walk an extra hundred feet from their Corollas. Stephen was just as angry at their weak spines: "If half these people had been in our units over there, we would have come home in body bags."

In public, I had framed my quixotic campaign as a fight to hold the assembly accountable, but honestly? Ethics reform was an extension of my deployment. I'd done anticorruption work in Afghanistan and heard a lot of my contacts say some version of this: "Afghanistan is not like America; all that matters in Afghan politics is money." Then I got to the Missouri legislature and found just one difference: nobody here worried about getting their head cut off. It was still all about money. And I found that offensive. I was offended on behalf of the soldiers still fighting to "spread democracy" or "defend democracy" and offended on behalf of a younger version of myself: the idealistic high schooler who skipped baseball games to attend debate tournaments or the college kid who never missed an airing of *The West Wing*.

If I could stamp out corruption in Missouri, maybe I could feel redeemed for not having done enough for my country in Afghanistan, and maybe I could make my country a little more like the place I'd been led to believe it was. With the help of a moderate Republican representative, I got the ethics bill passed on the final day of the session in 2010. I'd beaten the people who had opposed me, but what had I really won? Not much. The establishment went to work on eviscerating the bill in court before the governor had even signed it.

I won reelection that year, but I quickly recognized there were limits on what I could accomplish from my legislative office. Plus, I was antsy, feeling as though I'd been sitting in one spot too long. *I'd better get moving before it's too late*, I thought. Too late for what, I wasn't sure.

DIANA

This sense of hurry affected how Jason approached everything. It wasn't just how he saw his work. It was his general state of being. He was physically incapable of staying still.

If we'd go to a restaurant, as soon as I'd take my last bite of food, he'd stand up, ready to go. There was no leisurely after-dinner chit-chat. It was a big shift from how we used to enjoy meals out before Afghanistan.

When we were driving in the car, he'd get really anxious at stop-lights. If he was in the passenger seat, I could see his right foot pressing down on a phantom gas pedal. He needed to be moving. Staying still seemed to really agitate him.

We both still wanted to change the world. But his fight had become a lot more personal. I could tell that the load Jason was carrying on his shoulders was getting heavier and heavier. The thing is, he was putting it there himself. He thought it was up to him and him alone to pass ethics reform and save the state.

He was hard on the legislators and lobbyists who stood in his way, but he became harder and harder on himself. It was a side of him I had never seen before. The seventeen-year-old Jason Kander whom I met could do no wrong. He was confident and self-assured and filled with positive affirmations bestowed on him by his parents through his entire childhood. Now he started talking about himself in a tone that was different. That joy in being himself was fading. Nothing he did ever seemed to be enough.

I was worried that he was working himself too hard, but I didn't complain much. What he wanted to accomplish was important, and his passion for work was contagious. Who was I to slow him down?

In early 2011, I had started hearing rumors that Missouri's secretary of state, Robin Carnahan, might not run for reelection. This was a job I wanted very much—not only did the secretary of state's office oversee elections and voting rights, but it also crafted ballot language and was charged with financial regulation. I was doubting whether a legislator could have a lasting impact on ethics reform, but I knew that a secretary of state could. But getting the gig was a pretty serious long shot for me. I was a thirty-year-old progressive Democrat in Missouri without any statewide name recognition. It would be a slog.

When Robin finally announced that she would not seek reelection, I was ready. I had an email sitting in my drafts folder, announcing that I was considering a run for secretary of state.

You hear this language a lot from politicians who know perfectly well that they are running for something. We are just focused on serving our constituents. We are *considering* a campaign. We've formed an exploratory committee, which always sounds much more adventurous and outdoorsy than it is. It's not like anyone's fooled by this. "Exploring" actually means "I want to do all the things I usually do when running for office, like collecting endorsements and raising money to pay for a website and a staff and polling." But it also means "I don't yet want to be seen as a real candidate because I lack the funds to run a campaign or the knowledge to answer tough questions about the issues. And if it turns out that I can't raise enough money or the polling says I have no shot at the office, I'm better off as someone who 'explored' running rather than someone who tried and fell flat on his face."

I sat there for a good fifteen minutes, wondering if I should ditch all the BS and just come right out and say what I meant—"Charge the ball," as my dad said when he taught me to play shortstop. I called Chris Koster, who was then Missouri's attorney general, and he told me to just announce I was running.

"If you're willing to put yourself out there and be publicly humiliated in front of all your family and friends," he said, "it makes you dangerous."

I liked dangerous. I changed my email to declare that I was announcing my candidacy and hit send. Just like that I went from asking thirty-five thousand people to vote for me to asking six million.

But a few hours later I got some news that stopped me in my tracks. Mike Sanders, the well-established county executive in Kansas City, was himself considering a run. Mike and I were obviously known to each other, and just a year earlier, he had pledged to me that if Robin Carnahan didn't run for reelection, he'd give me my very first endorsement, and this had been a major factor in my decision to walk out onto this ledge.

This, by the way, is *exactly* why we do the "exploring a campaign" song and dance. If someone with a constituency twenty times bigger than yours, with hundreds of thousands of dollars sitting in a war chest, gets into the race, you can gracefully back out. Mike had been in Missouri politics since he was in his twenties, like I had, but he was fourteen years older and far more powerful.

Mike's relationships with Democratic kingmakers went much further back than mine, so they would feel obligated to back him. He'd also socked away huge piles of cash in political action committees with bullshit names like Integrity in Law Enforcement, which purported to be independent but were known to be controlled by Mike and his chief of staff. Meanwhile, I was a backbencher with no name recognition outside south Kansas City. I thought I'd been crashing into this campaign like Batman through a skylight. Now, instead of making a superhero landing, I felt exposed and vulnerable—like someone had snuck up behind me.

Suddenly, Mike wasn't just Mike anymore. He had become something much more threatening. I kept thinking of Hajji Zahir's surprise appearance that day at Sabet's office in 2006. I hardly slept at all the night after I heard that news. My night terrors were in full bloom. By morning I was as furious as I was sleep deprived. Mike Sanders had set a trap for me, and I'd walked right into it— unarmed. I phoned Mike several times, increasingly frantic, but he didn't pick up. The next day, as I made calls to ask for people's support, I found that nearly everyone in Kansas City said Mike had already called and made them promise not to commit until he made his decision.

It looked pretty hopeless. And if it had been anyone else, I might have bailed out, waited my turn at another office. Losing an election is bad; losing a primary can be a career-ender.

But, on the other hand, fuck that noise. I was so angry at Mike Sanders, I could taste it. He'd lied to my face. In fact, I wasn't just mad at him, I was mad at myself. I was an intelligence officer who had sat across from some of the most sinister and malicious liars this world had to offer, and I'd been able to spot their falsehoods through the textured glass of language translation. I was supposed to be good at smelling danger. I seemed to sense it everywhere else—when a stranger came up to me at a gas station; when something in the garage had been moved but not by me; when the house settled audibly at night; when someone wore a jacket on a warm day or walked directly behind me, outside my peripheral vision, on the street. My heart would balloon in my throat, my hands would become fists, and everything would start moving slowly, as if in water. My body had been trained to sniff out traps, but I had missed this one.

Then I took a breath. I was in my second term as a state representative, and I realized that although my anger on the house floor had felt good, it hadn't actually achieved much. I'd learned how much more I could accomplish when I made friends. Sitting in my colleague Jill Schupp's kitchen in St. Louis, I got up and began to pace and think. I tried to be objective and remove my own ambition from the equation—*Is it possible*, I asked myself, *that Mike would make a better candidate?* He'd certainly have more firepower in the primary.

But I knew in my gut that in the general election, when his advantages in fundraising and name recognition would evaporate, some of Mike's shady county dealings would make him a soft target for the Republicans. Sometimes, primaries are like college football, and the general election is the NFL—you might be the most feared quarterback in the Missouri Valley Conference, but the minute you reach the NFL, they rip your head off. Mike was going to lose, and Republicans would capture the secretary of state's office and do everything they could to demolish voting rights.

I knew I had at least a fighting chance to win the general. But first I had to win the primary—and I had to win it before it even began.

Mike was dodging my calls, but as luck would have it, we'd previously scheduled a meeting to catch up with each other, and it was just a couple of days away. I called his scheduler, who confirmed that the appointment was still on his calendar.

Mike had a permanent campaign team, and I didn't have so much as an intern. I knew he'd be "staffed" at our meeting and that it would be pretty hard to project strength if I showed up alone. So I called my friend Brian Noland, a first-year law student, and asked him for a favor.

Two days later, at a Walmart in Independence, Missouri, I parked next to Brian's white 2003 Chevy, got out of my car, and slipped into Brian's passenger seat. He chauffeured me the remaining three hundred feet to the Starbucks patio where Mike and I were to meet. Driving the candidate, carrying a notepad, and wearing a blazer, Brian looked every bit the part of a genuine campaign aide.

The meeting was scheduled to last an hour, so Mike took his time, beginning with a lengthy soliloquy about how hard it would be for him to say no to all the people asking him to do this thing for the sake of the party. He kept saying, "Jason, you know, it is what it is."

When it was my turn to speak, my tone was calm but firm. Ever since my deployment, I'd kept my cool in such situations—the more intense they got, the calmer I became. "I've talked to the same people you have. No one is asking you to run. You gave me your word. As a fellow soldier, I expect more of you. I'm very disappointed in you."

Mike had served in the army during the nineties, and that comment appeared to set him back a little, but only for a moment. He recovered, adjusted his position in his seat, and said, "Well, I've heard you've told some people you feel that way, and I feel bad about that."

"These aren't my feelings," I said. "These are facts."

"Well . . . it is what it is," he replied.

I don't like that expression. The only time I ever found it appropriate was when my grandmother died, and my grandfather kept saying it, over and over. It was his way of shrinking his grief down to a size where he could get a grip on it, understanding that what had happened had really happened and he was powerless to change it. But Mike wasn't powerless here. He was just full of shit.

"I'm not sure what you mean by that," I said. "But here's what you need to know. I'm running for secretary of state. That's happening. I'm sure you expected this to be a meeting that ends with me stepping aside. That's not happening. You can run if you want, but it's important for you to know you'll have to run against me. You start out with a ton of advantages, Mike, but you'll never outwork me, and I think you know it."

Mike looked pretty stunned. He was quiet. Rather than continue to speak, I looked him right in the eye and let my words hang in the air. I wanted him to know I was willing to lose, and that made me . . . dangerous.

He cleared his throat and forced an uncomfortable chuckle. "So I guess you didn't come here to discuss how we might talk about what's best for the party?"

"I know what's best for the party," I said, "and it's not you. You care about what's best for you. There's nothing to discuss. I'm running for secretary of state. You have to decide what you're doing."

"So you're saying I've got some thinking to do, huh?" Mike said, trying to sound casual.

"Sure. You can think about whether or not to be a man of your word."

"Well, like I said, it is what it is, so I'll have to think about what's next here."

"Great," I said, standing up and heading toward the parking lot without shaking his hand. "Call me in forty-eight hours and we'll talk about how to announce your endorsement."

Brian jogged along behind me to catch up while Mike and his guy sat still, looking stunned.

When we got into the car, my heart was trying to punch its way through my sternum—and I felt *amazing*. I was fully engaged, as though every part of me had been brought to life for the first time in years. I was flying a spaceship through an asteroid field, calmly and expertly punching every button and flipping every switch at the exact right moment while chaos reigned just outside the cockpit. That level of focus was a drug that had been damn near impossible to get stateside. I soaked in this hit for as long as I could.

"Well," laughed Brian—who hadn't needed to utter a word the entire meeting—"I'd say you stood your ground." He dropped me off in the Walmart parking lot and returned to class. The conversation with Mike had lasted six minutes.

For the next two days, I hustled like crazy to keep the pressure on Mike by securing and announcing every endorsement I could scrounge up. And right on schedule, two days later, Mike called to let me know he'd decided this wasn't the right race for him. I was alone in a hotel room at the Lake of the Ozarks, where the Teamsters Union was meeting and had invited me to speak. After I hung up, I raised my arms in triumph, flopped back on the bed in relief, and smiled. Somehow, on that Starbucks patio, I'd hustled my way to a primary win.

By the time I was running for secretary of state, I was in the final stages of my military service. I was as determined as ever to prove myself in politics because I was losing the part of me that allowed me to like myself.

Just a few months earlier, my reserve duty was still my escape. I lived for the opportunity to train the next generation of officers. I had made captain a year earlier, at which point my grandfather, a former private first class, told me, "Now you're a real officer."

I knew there would probably come a time when I would be deployed again, and I was honestly looking forward to it. I needed to go back. The eye twitch had disappeared, and I had managed to get the issues related to driving under control. I still had paralyzing nightmares, but they weren't always about the Taliban or Al Qaeda anymore, which I chalked up to progress. But that sick feeling I had, that feeling that every time I spoke about my service, my mouth was cashing a check I hadn't really earned—it was still there. Every time I said I was tired, or stressed, or red with rage at some political nonsense, I forced myself to remember how easy I'd had it compared to the guys I knew who were still over there. So what if I couldn't find a co-sponsor for a bill? At least I hadn't encountered any IEDs during my drive that morning. A reporter wrote something shitty about me—big deal. It wasn't like he'd put a gun to my head.

Then I got the call. I was getting deployed—to Kuwait.

Let me tell you about Kuwait. There was no war in Kuwait. The country was basically a set of US military bases connected by oil wells. The base I was going to had a Baskin-Robbins, a combination Pizza Hut–KFC, and a swimming pool. When I was in Afghanistan, I was out there collecting intelligence almost every day. I absolutely did not want to be sent away from my family and my career to spend a year at a desk *analyzing* that intel, making PowerPoint decks, and briefing colonels. Moreover, I'd have zero chance to finally shut up the stubborn voice in my head that told me I hadn't done enough.

I've never spoken publicly about what I did next. I got in the car and drove to Fort Leavenworth, which is just outside Kansas City, walked into the Center for Army Lessons Learned (CALL), and all but begged to be deployed back to Afghanistan. CALL's mission was to embed people in units of the Afghan National Army. Half the job was bringing back lessons from the field, and the other half was to act as an adviser to the ANA—which was not unlike the officer training I was doing now, though it carried a much higher likelihood of getting blown up or shot at.

Typically, CALL didn't take anyone below the rank of major or lieutenant colonel, but I pulled out every persuasion strategy I had to get them to make an exception, and they agreed to take me as a captain. I would be sent on a nine-month deployment to Kandahar, which, in 2011, was tied with Helmand province (its neighbor) for most dangerous and violent place on the planet.

I was elated. I was finally going back, and this time I wouldn't merely be going to meetings along with Salam and collecting information. I would be kicking in doors with Afghan troops and collecting bad guys. The folks at CALL warned me that "green on blue attacks" (Afghan forces turning on their coalition trainers) were at this time the worst they had ever been, and the problem was especially great in Kandahar. But I was undeterred.

Now I could fix whatever was wrong with me. This—*this*—would surely be enough, I told myself. If I could go back and make a real impact, maybe even get myself hurt. Then maybe I'd finally feel like I'd really done something.

Diana

It didn't make any sense, of course. Why would he want to go back?

Years earlier, when Jason told me he was going to Afghanistan, my reply was "Then I'm buying a motorcycle." I figured I could make any demand I wanted at that moment, so I went for the thing he would never agree to before Afghanistan. It's not that he didn't want me to spend money on a motorcycle. He was just worried about my safety.

Jason's first deployment had been awful for me. Between the loneliness and the relentless fear for his life, I'd learned to distract myself in any way that I could. I bought the motorcycle, worked about fourteen hours a day, applied to be on reality TV shows, and when all of that failed to be an adequate distraction, I took up mixed martial arts and began to train to fight.

As hard as all that had been, though, I didn't feel like putting up a fight against the second deployment. Jason kept saying he needed the closure, and I wanted to support him in that. My plan was to hang on and get through it.

I had everything lined up, down to the date I was shipping out. All that remained was a simple formality—my regimental commander had to sign a document releasing me back to active duty. I handed it to him at drill weekend and, with great pride and enthusiasm, told him about the adventure I had in front of me. And then he refused to sign. He told me he needed the adjutant general, the two-star in charge of the entire Missouri National Guard, to give him permission.

"Sir," I said, "this is your authority. You don't need anyone's permission."

"Yes," he said, "but there's no harm in asking."

But I knew there could be harm in asking. I felt my chance at redemption slipping away.

"Sir, I deal with the general in the legislature a lot. If you elevate it to him, he's going to worry that it looks like a political favor to me, and he'll have no choice—he'll have to say no."

The man looked at me, clearly wondering, *Who the hell is this random captain telling me how to do my job and acting like he's pals with my boss's boss's boss?*

"I'll call his office on Monday," he said dismissively. "We'll see what he says."

After I tried again to make my case, asking him to just exercise his authority, he let me know he just wouldn't do that.

Of course, he called the general, who said exactly what I knew he'd say—that it would have been fine to release me, but now that the matter had risen to his level, it might look like I got special treatment, and so his hands were tied.

I was *furious*. I was sure I'd been on the verge of fixing everything that was wrong with me, and now I was staring at nine months, at least, of writing memos in windowless rooms, far from the people I loved, in a place where the only action I was likely to see was a game of pickup football.

Shortly after this disappointment, while doing a standard paperwork review with my unit's readiness sergeant, I discovered that my contract with the army was almost up. I wasn't the type of person—or lawyer—who read the fine print, and I'd always assumed my eight-year commitment began the day I got my commission, but it had actually started two years earlier, when I'd enlisted. Instead of two years away, the end of my contract was two months away.

It was disorienting. I'd always planned to be in the service for a minimum of twenty years. I wanted to make lieutenant colonel—at least.

But now I had to make a choice. Stay in and go to Kuwait, only to come home and "fly a desk" on weekends at brigade or battalion staff for ten more years, or give up the only world where I felt completely at ease. No part of me wanted to leave the military. That life made sense to me. It wasn't just that I loved the breakfasts and the training and the camaraderie. It was the awareness of purpose and the order. Every day I was a soldier was a day I woke up and I knew exactly what I was doing and why I was doing it. I put on the same clothes, I knew who my boss was, I knew who I was responsible for, and I knew my mission. Truth be told, ever since I'd gotten back from my deployment, everything I'd done in the civilian world had been a quest for that level of order and meaning. And nothing—not my work in the legislature, not my role in court as a plaintiff's attorney—had come close.

My sense of identity began and ended with my connection to the army. Years earlier, when I'd first joined ROTC, I had felt like a law student who also did some army training. Now I was a lawyer and a politician, but I felt like a soldier who happened to do a lot of legal and political work.

I didn't want to give up my commission, so I figured I may as well see if there was a way to make the Kuwait deployment less boring. I went to see the leadership of the Missouri National Guard's Judge Advocate General's Corps (the JAG Corps). I thought intelligence work in Kuwait would be dull, but with all that idle time on their hands, soldiers would find ways to get into trouble. If I could go as a JAG (a lawyer in the army), I'd at least be busy, and maybe I could prevent a few privates from throwing away their careers.

In my meeting with the JAG Corps, the colonel in charge confirmed that I could become a JAG by attending the ten-week army law school in Virginia, but first I had to go through the Basic Officer's Leadership Course (BOLC). This was strange, because the colonel knew I had been teaching that course for the previous three years.

This colonel was literally saying I had to retake a course for which I was lead trainer at the Missouri National Guard. I'd been frank with him at the beginning of our conversation. He knew I was considering whether to stay in the service past my commitment. He understood that this was a chance for the army to retain a captain with intelligence training, a top-secret security clearance, certifications as an instructor, a deployment under his belt, and a law degree from Georgetown. Heck, just a year earlier I'd been a top-ten national finalist for a reserve junior officer of the year award.

He acknowledged the BOLC requirement was waivable but said he wouldn't support my application for a waiver. So, no way was I getting one.

I had always heard the army had a bad habit of undervaluing company-grade officers, meaning captains and below, but this was the first time I experienced it firsthand.

By the time I left the JAG office, I was heartbroken. I felt like the army had just broken up with me. I loved the army so much, and this felt like it didn't give a shit about me. The only place its leaders were willing to send me was Kuwait, which wasn't going to achieve what I was looking for. If this was the only thing they could use me for, clearly they didn't need me. In 2011, a few weeks after that meeting with Mike Sanders, I gave in and resigned my commission.

My last drill weekend was spent out-processing at state headquarters. I knew I was facing my final time in uniform, but I put off writing the actual resignation memo—just a few sentences—until the last moment. Then I turned it in, along with a couple of duffels' worth of equipment, signed a statement affirming that I had not been sexually assaulted, and left the building. None of the people I'd actually served with were present. The process felt stale, bureaucratic, unceremonious, and lonesome.

As I walked across the parking lot, headed for my 2008 Ford Focus, I saw a soldier walking toward the building. *This will probably be my final salute*, I thought. I didn't know the man, but I could see his rank. He was a warrant officer, so I'd be returning his salute. My mind flashed to nine years earlier, when I crossed the Georgetown parking lot and rendered my first salute to a senior cadet. Now, a few paces apart, the warrant officer and I made eye contact as his index finger touched the brim of his soft cap and he said, "Hooah, sir." I savored the moment for a split second before returning the salute and the "Hooah." Then we strode past each other and I thought, *He has no idea I'm about to be a civilian.* The finality of it floated into my soul on a cloud of shame.

I spent the two-and-a-half-hour drive home trying to understand who I was now, to rewrite my story, create a narrative in which I could live. I couldn't come up with one. I settled on the idea that I'd probably end up back in the service at some point. This was a break, not a breakup. But deep down, I knew: it was over.

When I got home and undressed, I realized that my soft cap and my boots could go into storage, rather than into the closet of the spare bedroom, where my uniforms and gear had lived for years. I thought about all the times Diana had pulled my boots off for me back in law school. Back at the beginning.

But I couldn't bring myself to do this. I placed the cap and boots right back in their usual spot and decided that when my uniform was clean, I'd hang it where it always had hung. I was reassuring myself that I'd return.

Something happened when I took off that uniform for the last time. Something terrible. The part of me that I liked and respected the most died.

It had only been a few months since my commander had refused to sign that form to release me to active duty. In a short time I went from being sure that I'd finally be able to fix what was wrong with me to no longer being an officer in the US Army. And now I was a civilian. I sat down on the bed and wondered if I should cry, but I couldn't. It had been a very long time since I'd been able to cry. I was deeply sad, but I felt I had no right to mourn: I had survived Afghanistan, had done only one four-month tour, during which nothing bad had happened to me. Why didn't I feel lucky? Or grateful? What possible right did I have to mourn anything? There were so many soldiers who didn't make it back . . .

T.J. earned the right to leave this way, I thought, *but I didn't.*

I didn't have the first idea of how to process my own sadness, so I turned to a reliable old friend: anger.

I should be in Kandahar right now, I thought, *and I'm still pissed about it.*

5

I SHOULD BE BETTER BY NOW

I felt like I was going to die. Like death was *happening*. Now. Here. With aching knuckles I was gripping the underside of my chair, like it was going to blast off. I could see my knees, but I couldn't feel them; it was as if they'd dissolved. I was frozen in place, blood pounding in my ears, willing myself not to die. If I let go of the chair, I would die. Every neuron in my brain was screaming one message: *survive!* I knew this feeling of utter helplessness. I'd felt it while lying prone behind the rock wall in Jalalabad, hearing the staccato of small arms fire and the occasional boom of a grenade. My mind summoned the memory so intensely, I could smell the grass and dirt under my nose.

But this wasn't Jalalabad. This was a suite at the Q Hotel in Kansas City. There was a minibar. The only booms came from my phone, vibrating on the other side of the room, signaling that I had dozens of text messages. I was in the far corner of the room, facing the door. Diana was sitting on the bed, distracting herself by half-watching the movie *Ted*, something we'd planned to do, but I'd made it through only the first five minutes. Then I'd strapped myself into this chair. Mostly, Diana was talking to Abe Rakov, my campaign manager, who was sitting with us, bouncing his laptop nervously on his knees, the screen turned toward him so that I'd know only what he told me. Down the street, at Californos Bar in Westport, the watch party was in full swing. It was election night 2012, and I was losing.

Nothing I did could shut off the feeling of sudden agony. This made no sense to me: my entire *thing* had become that the more intense the situation, the more everyone else's hair was on fire, the calmer I was. That was how my body had learned to channel adrenaline—in fact, I felt comfortable when that was happening. But that election night, there was nothing left to do but sit and wait. It was just me, the chair, and my brain, and it was unbearable. Abe double-tapped the touchpad of his laptop every fifteen seconds, refreshing the live feed of election returns. Every few minutes he'd call out the latest numbers, which were mostly bad news, so he'd soften the blow, saying "It's early" or "Still a long way to go." Every time a new batch of votes was posted, I steeled myself, the way I'd forced myself to peer through the opening in that rock wall, hoping that one of the two dozen barefoot, unarmed Afghan recruits hadn't sold us out to the Taliban. This same thing happened occasionally when I had no control over a stressful situation—my mind would return to a memory from Afghanistan. It happened a lot when I was running for secretary of state.

The voter outreach playbook I'd followed in my first campaigns was worthless now: I couldn't knock on every door this time. And unlike Mike Sanders, I hadn't amassed a gigantic war chest, with slush funds up to my ears. I'd had to beg and plead and wheedle and flatter for every single dollar. But even Mike's money paled in comparison to what my Republican opponent had: a billionaire sugar daddy.

Technically, my opponent was Missouri's speaker of the house pro tempore, Shane Schoeller. But as soon as he got into the race, Shane became the avatar of a St. Louis billionaire who belonged to that distinctly American species of very rich guys whose wealth permits them to make a hobby out of governing from home. Think of the Kochs and the late Sheldon Adelson, who owned the entire slate of Republican presidential candidates. This guy had spotted a bargain: for a fraction of what it cost to own, say, a GOP senator from Florida, he could buy himself a secretary of state.

While individual donations to federal campaigns are capped at $2,800 per candidate, Missouri had no limit on donations to statewide candidates. So there was nothing to stop one rich guy from dropping hundreds of thousands of dollars into Shane's campaign account, whether directly or laundered through a Republican PAC (which he did, on three separate occasions)—and I couldn't do a damn thing about it.

Here's the thing about running for office. It's easy to assume, based on how it's portrayed on TV, that kissing babies and marching in parades and giving speeches and making TV ads are the heart and soul of it. The grim reality is that at least 90 percent of a campaign is made up of Call Time: parking yourself in a little room with a phone and a laptop and begging people to give you $200, and if they can do $200, could they make it $250?

All I could do was the only thing that took away that gnawing sense that I was a fraud: work myself to death.

Fortunately, I had a secret weapon: a fresh-faced kid from Northwestern named Abe Rakov.

Stephen Webber had introduced me to Abe, who had helped put together the strategy for Stephen's statehouse campaign, and he'd done it brilliantly. Other than that, Abe had no ties to Missouri. Born in California, raised first in New Mexico and then in the Florida Keys, Abe had been college roommates with Stephen's brother. We both knew we couldn't let a talent like Abe leave the state. In places like Missouri, we're always conscious of how hard it can be to keep brilliant people from being drawn to the big cities on the coasts—it's especially personal for me, since I grew up watching damn near every Royals player I fell in love with inevitably get shipped off to the hated Yankees. The very first time she met Abe, Diana said to me: "When you run statewide, we have to convince that guy to manage the campaign."

Diana and I used to joke that Abe was like Buddha: he always seemed to know the answer, even when he presumably had no relevant experience or frame of reference. We were young in this political game, and he was five years younger than us, so how did he always know what to do? Thus our theory: he just went and sat under a tree until the answer became clear.

Fortunately, I was able to convince him to run the campaign. I'd been raising money and traveling the state for five months—with a skeleton staff and limited success—when Abe joined us on the road in March 2012. And the roads in Missouri are long, flat, and empty. We'd be gone for days at a time, crashing in supporters' guest rooms, making fundraising calls from the car, shaking hands at every county fair or slightly busy public gathering we could find. In the evenings, we'd sit through long, boring local events just for the privilege of giving a three-to-five-minute speech while half the room was in the buffet line. And honestly, I loved it. Abe's used Ford Escape, wrapped in our campaign logo, ferried us—Abe, Kellyn Sloan, Kyle Juvers, Diana, and me—as we hollered along to "Wagon Wheel" by Old Crow Medicine Show, which we played on repeat. We spent so much time on the road that Diana was able to write a whole book, her first one, in the back of that Ford Escape.

Abe and I developed a deep closeness, the kind that rarely exists between a politician and a campaign manager. We were together constantly, sharing rooms on the road, living every insane moment together. A few months into the campaign, I asked him to be my chief of staff if I won, and the offer didn't come as a surprise. By then, we'd become best friends.

Back at campaign headquarters, Shawn Borich, Liz Zerr, and a squad of interns mined the internet for prospective donors and begged existing ones to host fundraisers while Kevin Flannery tried to convince activists across Missouri to care about the race for secretary of state.

Team Kander had grown, and we were a tight bunch. My friend Chris Koster, who was running for reelection as Missouri's attorney general, described it best on the frigid early morning of the Mizzou Homecoming parade. After watching Liz lead us in a chant of "We believe that we will win!" Chris said, "My God, Kander, that's not a campaign, that's a cult."

For the first time since Afghanistan, I was on a big adventure, doing meaningful work alongside people I liked.

When I got into the race, the outgoing secretary of state, Robin

Carnahan, gave me some advice. "This is a horse race," she said, "and *you* are the horse. You need to understand that you are going to have all these people working for you, all pulling your reins in different directions. You are in charge of the horse department—because no one else is. No one else will think about the horse. You have to make time for yourself and your family. You have to take care of the horse."

It was great advice, and I planned to completely ignore it. Running for statewide office was going to push me to my limit, and well past it. That was one of its main selling points. I wanted—*needed*—to be so busy, I'd barely have time to think. During my first campaign, when I was struggling with anger and survivor's guilt, every door I knocked on, every yard sign commitment I scored, gave me a tiny but desperately needed whisper of relief. In the legislature, staying late to answer every single email—which I'd finally stopped doing—had kept me from having to go home to sleep. As long as I was doing stuff, pushing for every dollar and every vote, I felt good—in fact it was the only time I did. If, at the end of the day, I was completely hollowed out by exhaustion, my eyes stinging with fatigue, my back a spiral of pain, then—only then—did I feel I had done well. But in moments of stillness, when I was alone with my thoughts, or as night fell, things had started to get dark.

DIANA

Jason disregarded Robin's advice, and frankly so did I. We were trying to change the world. That kind of mission takes a lot of sacrifice. Once it was over, once Jason had won, then maybe we could slow down. But right then, we were asking everyone around us to give everything they had. How could we possibly ask less of ourselves?

For Jason, taking care of himself meant doing whatever would allow his body and mind to do the work that had to be done—like a football player getting a painkiller shot in the locker room at halftime, in order to get back out on the field. That's the level of commitment we both had to have.

When Jason was in Afghanistan, I was so incredibly sad and scared that I could barely get anything done. I tried all kinds of activities that could offer me that kind of painkiller, and ironically, the one that seemed to do the trick came with a fair amount of pain. I fell in love with mixed martial arts because it's impossible to be sad while you're trying not to get punched in the face. Fighting was all-consuming, and I desperately needed what it did for me. I'd pull into the gym's parking lot, crying my eyes out, my brain a swirl of overwhelming thoughts. Then I'd come out feeling okay enough to be productive for a few hours. To me, that was taking care of myself. So I saw what was going on with Jason, but I also understood it.

Instead of listening to Robin's advice, I launched my campaign at full gallop and vowed not to slow down. When people cautioned me that a campaign "is a marathon and not a sprint," I would reply that it's a lot easier to win if you sprint the whole marathon. Fund-raising was a volume game, especially since I was going up against a billionaire, so I didn't let up. Because knocking on doors wasn't going to cut it statewide, dialing for dollars was the only way I could physically outpace the competition. Whether I was calling from our office, or more likely from the passenger seat of Abe's car as we traveled around the state, I viewed it all as a test of endurance.

Sometimes, late at night when we couldn't make any more calls, I'd nap in the car. After waking up suddenly, with a scream or a gasp for air because I'd just experienced a night terror, I'd self-consciously apologize to Abe, and he'd generously act like he hadn't noticed anything.

I just kept marching.

The army had given me that superpower: I could endure suffering. I ate shitty food, my back pain got worse. I even tried getting really drunk every couple of months, like the time I—in my official role—attended the deployment ceremony for the soldiers going to Kuwait. Even though I expected all of them to come home safely,

I couldn't help but feel deep, stinging shame to be there as a self-promoting politician and not a departing soldier. Abe sensed that something was wrong. As soon as the event was over, instead of having me go right back to fundraising calls like usual, he said, "Let's go buy some cowboy boots and then get drunk." And so we did.

He was trying to make me feel better, like any good friend would do. Somebody had to try to take care of the horse. He knew how much talking baseball calmed me, so he became as knowledgeable about the Royals farm system as I was—a stretch for a Dodgers fan. He was also the guy reminding me to call Diana whenever we had downtime. He didn't just take care of me; he took care of my family—and did a better job of it than I did.

Around 11:15 on election night, I couldn't take it anymore. I wanted to stop feeling helpless and vulnerable more than I wanted to win, so I told Abe I was ready to call my opponent and concede. Abe ignored me, but I insisted. Abe was not having it. It was the only time he ever yelled at me. "Don't be stupid!" he said. "There are still enough votes out there for us to win!" He refused to give me Shane Schoeller's number, but I couldn't stay in that room. I had been sitting in one place too long. Time to move.

It was a five-minute walk to the watch party at Californos, and the fresh air, combined with the fact that Abe's laptop had no Wi-Fi access outdoors, gave me momentary relief from the dread of imminent death. Then, once we were in the holding room on the edge of the party, with close friends and family, my instincts as a performer took over, and the panic subsided. I had something to focus on, something I could control: projecting calm and confidence for my parents and my brothers.

At around 11:30 p.m., we learned that President Obama was about to give his victory speech, so we filed into the main room to watch. The president, cool as always, strode onto stage with Mrs. Obama and their girls, a gorgeous symbol of America's journey. The room erupted in cheers and hugging as my political hero began to speak, but after the first few lines, I didn't hear any of it.

Left with nothing to control but my own brain, I floated between two scenes: standing in the back of my own watch party and sitting in the passenger seat of a Mitsubishi Pajero, dangerously stuck in Kabul traffic. The face of a little Afghan boy flashed in my mind, his terrified expression framed by the iron sight of my M-16.

He had jumped onto the driver's side of our vehicle at a time when suicide bombers had recently taken to flinging themselves into convoys and holding on to a vehicle long enough to detonate the bomb. I'd responded instinctively, raising my rifle to kill him before he took us with him. My dominant eye met both of his as I looked down the barrel. Thankfully the message *HE'S JUST A KID* raced from my ocular nerve to my frontal lobe just in time to abort the near-automatic process of squeezing the trigger.

It wasn't shocking to see his face here, while waiting for our election results. There was at least one moment every day when it flashed before me, sometimes sticking around long enough to make me wonder how the little boy's encounter with me might have changed the course of his life, or to make me physically ill as I pondered what could have happened instead. I might also feel guilt for every soldier who'd been in that situation and—for whatever reason—hadn't been able to see clearly enough to avoid firing.

The final batch of returns from Kansas City had come in, shrinking Shane Schoeller's lead, and all that was left was north St. Louis. My campaign team knew this could give us a boost, but we didn't know how much.

At around midnight, a few minutes after Obama wrapped up and the talking heads took over on the big screen, we suddenly pulled ahead by twenty-eight hundred votes. The crowd at Californos went wild with joy, cheering and hugging and crying.

I felt better, but the joy everyone else was expressing was beyond my grasp. Winning didn't feel like I thought it would. Countless times since getting into the race, I'd imagined what this moment would be like, and yes, the feeling would be elation. It would be like travel ball years ago, when I'd come in as a late-inning reliever in some little town in Kansas or Iowa or Nebraska and get a clutch out in a close game. I remembered that feeling—that joy. But I couldn't feel it now. What I experienced was relief. I wasn't going to die tonight.

A few days later, Diana and I flew to northern California, rented a car, and spent eight days driving down the Pacific Coast Highway. It was our first true vacation in years.

My reward for winning the election—other than not dying—was something I wanted more than anything else: a bonanza of obligations to occupy my time and my mind. I now had a staff of 250 people. I was in charge of safeguarding Missourians' voting rights, writing ballot language, investigating financial crimes, and fulfilling a bunch of other responsibilities you may have no idea your secretary of state handles. It was the perfect role for someone desperate to avoid dealing with himself.

My first full day in that role began in the parking lot outside the office. Abe, whom I made my chief of staff, planned to carpool to work with me each day. He mentioned a recent *Huffington Post* article profiling me and labeling me the first millennial in the country to hold statewide office. "I'm only going to bring this up once, and we won't have to talk about it again for several years," he said. "At this moment, out of every American under the age of thirty-two, you have the best chance of becoming president. So let's try not to fuck this up."

"Good plan," I said.

At that time the term "millennial" was getting tossed around like an insult (how we were "killing" some core American value or how lazy we supposedly were), I leaned hard into my new branding as the first millennial to hold statewide office. The attention it brought me was a narcotic. I remember the moment I found out my Twitter account had been verified, which may as well have been called validation instead of verification. In those early days, it wasn't enough just to hold an office or be a staff writer at a website—it was a more subjective test that seemed to come down to "being someone who mattered." I had been the guy who eschewed honorifics, but suddenly I was chasing this symbol. To me, that little blue check mark was tantamount to American knighthood.

I said yes to every media request I could. When I looked into a camera, with the stage lights warm on my face, a news anchor's voice in my earpiece, and heard my own voice pour warmly from me as I did the thing I *knew* I did better than almost anyone—talk—a rush of positive neurotransmitters flowed through me, and it could last for hours. I'd have this lovely afterglow when I finished a really good interview—it was, honestly, a little like sex.

If media attention was one kind of distraction, work was the other. I hurled myself into projects and worked until I was so destroyed, I could barely move.

I had a new enemy in my sights: the rising movement across America to suppress voting rights, and specifically its advocates in Missouri. "Movement" isn't even the right term. Calling this effort a movement implies that large groups of people were clamoring to suppress the vote. They weren't. Instead, it was a very small group of incredibly wealthy people.

Republican politicians and their corporate funders looked at the changing demographics of America and saw a problem: Black and brown people, as well as the vast majority of the millennial generation, were decidedly not in their camp. And each of those groups was becoming an increasingly big part of the American electorate. If the politicians sought to earn these people's votes by fighting climate change or tackling income inequality, they'd lose boatloads of corporate campaign cash; if instead they made this effort by prioritizing racial justice or embracing common-sense gun reforms, truckloads of reliable rural white voters would stay home.

So they were in a pickle, and thus they chose to win by subtraction. Rather than try to win over Black and brown people or young people, they decided to keep them from voting. Today, thanks to the efforts of people like Stacey Abrams, we know all about voter suppression, but in 2013, when I became secretary of state, only those of us fighting it even knew the term, and even the majority of those in my own party believed in the myth of voter fraud.

The only chance I had of beating back voter suppression in Missouri was to go on the offense. I had to educate the public until they no longer disagreed with me.

Abe and I put our whole team to work, using the bully pulpit of my office to attack voter suppression bills, like Photo ID, as publicly as possible. Despite a veto-proof Republican supermajority in the state legislature, we mostly held back the newly proposed voter suppression laws.

But with my higher profile, a new kind of trouble began brewing.

My nightmares evolved. Instead of being in a meeting with an Afghan warlord, I was at home. Instead of coming to attack me, violent people were coming for Diana, or my parents, or my younger brother, and I couldn't stop them. It would be the middle of the night and someone would be at the door, and they'd overpower me and rush past. I'd wake up in terror—but I couldn't move a muscle because my body remained asleep. Every night Diana would wake up as I struggled, grunting to signal to her. She would sit bolt upright and begin rolling my body back and forth. After a few seconds, which felt like long minutes to me, her jostling me would interrupt the paralysis, allowing me to sit up in bed. I'd suck in a giant breath as though I'd emerged from a near drowning.

DIANA

When you live with someone who constantly tells you how dangerous the world is, how vulnerable you and your family are, you start to believe it.

Every night Jason would wake up, and so would I, and he'd proceed to recount every dream in detail. In the middle of the night, suddenly it was Horrible Story Time. He would provide very vivid details. And when you're half-asleep like I was, you're very impressionable. The effects of this experience soaked into me slowly. I didn't have nightmares. Instead, my days started to feature jarringly precise visions of terrible things that might happen. I began to develop severe anxiety, manifested in these daymares (nightmares that occur while you're awake).

One time I was stranded in an airport, at a terminal filled with angry travelers. I was standing in line, waiting to rebook my flight, and the man behind me remarked, "You're in an awfully good mood for someone stuck in an airport."

"Just got a positive outlook, I guess," I replied. But I remember think-ing, You have no idea, sir. I've had visions all day of my flight go-ing down, the panic of passengers who will never see their loved ones again, and the feeling of hot fire melting my skin as the plane bursts into flames. *I had to chuckle to myself at the horror of it all.*

This phenomenon was brand-new to me, but I assumed it had to do with Jason's rising public profile. It was great that people had started to recognize him, which meant it would be easier to accomplish the big things he was working on. But at night all that recognition was transformed into a terrifying threat.

When Jason was on the road, I would stay up late into the night with panic attacks, expecting a home intrusion. As in, I was staring at the door, just waiting for it to open. I'd go over different scenarios in my mind, making sure that my body had the muscle memory to defend itself if suddenly jarred from sleep. I had trouble taking full breaths, and eventually, tears would start pouring out of me. This was annoying because it made it hard to hear if anyone was in the house. I'd eventually pass out, clutching a knife under my pillow and hoping I had the strength to use it.

I struggled with extreme fatigue—I took every blood test and tried every diet change I could find, trying to diagnose the problem. I be-came very angry. I wore an Incredible Hulk bracelet to remind myself to stay calm. It wasn't very effective. And I developed the obsessive eye for danger that Jason had had since Afghanistan. His compulsive safety precautions didn't seem weird or over the top. I was desperate for more of them.

We knew other couples weren't doing this. But we weren't other couples.

The nightmares (and literally any noise in the house, whether real or imagined) would propel me out of bed to patrol our home, my mind wild, expecting danger. I would move along the walls, a .357 revolver in hand, clearing the house room by room, just as I'd been trained to do. The revolver had been my father's off-duty weapon as a police officer, and he'd given it to me a few years earlier. After I'd served a few months as secretary of state, Diana made a persuasive argument against keeping a gun in the house, and I reluctantly gave the pistol to my father-in-law to store in his safe. The night patrols continued unabated, however. A hammer replaced the .357.

I wasn't stupid. I knew something was wrong with me. I'd known it for a while. Normal people weren't afraid of going to sleep; they didn't prowl their houses every night; they didn't bar their spouse from answering the door if they weren't home. And what was most bewildering and frustrating was that I knew I *should* be happy. I had won my political race. I'd started getting national media attention. And more than that, we'd moved to a house in the town of Columbia, midway between Kansas City and Jefferson City, which gave Diana and me something we'd never had before: stability. For the previous decade, I'd been bouncing from DC to Fort Huachuca, then to Kabul, then to Kansas City, then spending half my time in Jefferson City. Over the past year, I'd pretty much lived on interstates 70, 35, and 44. During that decade I'd been a student, a soldier, an officer, a corporate lawyer, a teacher, a civil justice lawyer, a legislator, and—not nearly as much as I should have been—a husband.

And somewhere along the Pacific Coast Highway, we had added another job to that list: father.

Our son, True Steven Kander, was born in 2013.

Like any father, I was in love immediately. And as is the case for some fathers, that love was interwoven with fear. Every dad feels that evolutionary protective instinct toward his kids, but mine was something more.

For as long as I can remember, I've felt some level of protective instinct. I'd learned it—in the best way possible—from my parents. The army saw that instinct and said, "Oh, let's whip this up into something *really* strong." Going to Afghanistan basically set it on fire. But having a baby? It was like pouring jet fuel on it.

I was obsessed with SIDS (sudden infant death syndrome). From the day we brought True home, I would lie all night next to him as he rested in his bassinet. I barely breathed, so I could hear *him* breathing. I'd fight off sleep until I was so exhausted, I passed out. When he woke up and started crying, I felt relieved.

When he eventually started sleeping in his own room six feet down the hall from ours, things got worse: I'd lie awake in bed, staring down that hallway for any hint of a shadow. People knew where we lived, and I didn't trust our house alarm to alert me if an intruder managed to sneak in. Any creak made by the house settling, any squirrel in the gutters, and I'd be wide awake and ready to fight.

I became convinced that people were plotting to kidnap my son. *Convinced.* I didn't suspect it—I knew it. Some nights it wasn't enough to stare down that hallway; all I could think about was the window in True's room that faced the front yard. In order to calm myself enough to get some much-needed sleep, I'd take my pillow down the hall to his room, shut the door behind me, put the Diaper Genie in front of his door (so that it'd make a noise if someone opened it), and lie down to sleep on the floor between the window and his bed. That way, anyone coming in through the window in the dark would step on me.

And when I could sleep, it was no reprieve. From the day True was born, he entered the cast of characters in my nightmares, people I inevitably failed to protect.

But then, eventually, day would break, the night terrors would dissolve, and I would remember to feel grateful for everything I had. I loved my family, I loved my job, and I was, on the whole, a delightful person to be around. I started every day in the office by visiting any employee who was celebrating a birthday. I'd find them at their desk, pull up a chair, and spend a few minutes learning about that person's job or talking about their family photos. It was the perfect way to start a day.

I thought I was okay. I thought I was handling it.

6

YOUNG MAN IN A HURRY

You won't hear this from a lot of politicians, but hands down, one of the best things about the job is how much it involves knowing and keeping secrets. In this regard, it's a lot like intelligence work, only most of the time the secrets aren't a matter of life and death. Instead they concern the vast gulf between a person's public statements and their private opinions. For instance, a statewide politician who spent a lot on polling once said to me, "I'm like the CIA of political data, man. I know what everybody really thinks." And just like the situation in the intel world, the higher you rise, the cooler the secrets you get to be a part of.

On February 19, 2015, I was in a great mood because I had been keeping a secret for weeks, and today I would get to reveal it. I parked my car in a lot in downtown Columbia, MO, zipped up my coat, and stuck in my earbuds. The walk to the office I was heading for was short, only a quarter mile, but it gave me enough time to put on "Uptown Funk" and pretty much go dancing down the sidewalk. I felt more excited than I had in a long, long time because I knew I was about to shock the hell out of a lot of people.

The office was upstairs from a law firm and a diner on Walnut Street. Only a few people were in: Abe, Kellyn Sloan, who'd been my call-time manager in the secretary of state race, and a couple of visitors from Washington. Abe and Kellyn had just resigned their posts in the secretary of state's office.

A few years earlier, Diana had been part-owner of a new bar in Kansas City, and I remembered how it felt on opening night, in those last few minutes before the place opened to the public for the very first time—I recalled the tension, the excitement, the fear that no one would even care.

Outside a very tight-knit circle, only three other people knew what those of us at this office were about to do: US senators Chuck Schumer, Claire McCaskill, and Jon Tester. We'd even shot an announcement video, in secret, at the Kansas City airport, at an empty gate in a terminal that was scheduled for demolition. At 9 a.m. that morning, Abe hit send on a press release and posted a video that broke open the secret: I was officially running for the US Senate. I was thirty-three years old. The office in Columbia was campaign headquarters.

For once, it hadn't been my idea. Abe and Stephen, it turned out, had been conspiring to subtly convince me that (1) I could run and (2) I could win. I'd been hesitant. When Abe brought it up, my first reaction was "Man, I'm in this new job, I have a new baby, I finally have a house with furniture, and I actually sleep in the same bed every night." Sure, I was still prowling the bushes every night before I got into that bed, but I was handling it. I thought life was actually good, or as good as I deserved. I wanted to believe the itch that had driven me was soothed, at least for now. I told myself I wanted to get reelected as secretary of state a couple of times and see where things went from there. I had to tell myself this a lot, though.

One day, Abe convinced me to do just one poll. It was what we call a "benchmark poll," which starts by hypothesizing a head-to-head matchup, which in this case included me, a relative unknown, and Senator Roy Blunt, who had been serving in elected office for nine years when I was *born*. At the start of the polling call, the person phoning asks where the voter stands. Then they say, "Let me tell you about Jason Kander; here are his positives." Then they discuss Roy Blunt and his positives. Then they cover the negatives on each, and in this way they basically play out the whole campaign by giving voters the information they would have on election day. One call takes about forty-five minutes. The final question: "So, did any of that change your mind?"

Abe was at Chipotle when he got the results, and when he returned to the office he said, "I decided to pay for my burrito *and* yours." The poll had me up by a hair.

Now we knew something no one else knew: I could win.

Whenever I thought about this amazing secret, I pictured the sheer firepower a race for the US Senate would bring to the fight going on in my mind. This, *this* might finally be enough. First, I was staring at a two-year campaign that would consume me completely. But more important, as a US senator, I could really help people—for one thing, virtually no one in the Senate had served in Afghanistan. If I could speak for other Afghanistan veterans on a national stage, surely that would bring me some sense of redemption, right?

And now that I knew I could win . . . well, now I sort of had to.

The minute the announcement went out, we hit the phones, trying to raise as much money as we could. All morning, in between my own calls, I kept badgering Kellyn, telling her, "They need to call this person, we need to be double-dialing, the staff need to be getting us more call sheets. Are we responsible for follow-up, or are they doing it?" Finally, a few hours in, I started to say, "Hey, can you tell them we—"

"It's just me!" she cried. "You keep saying 'they' and 'we,' but there's no they! It's just me over here! There's nobody else!"

I stopped and put the phone down. I looked around the room, which was just as small as the call room we'd used four years earlier. Kellyn had a desk, a computer, and a chair, and I had one of those nylon foldout chairs that parents bring to their kid's soccer games. There was nothing on the walls. Kellyn and I had been working together since she was a high schooler volunteering on my state rep race, and this was the first time she'd ever gotten mad at me. She was right. There was no team—they hadn't moved to Missouri yet. I apologized profusely.

Within an hour of my announcement, the Missouri GOP released a statement blasting me as overambitious.

Reader, let's talk about ambition. First of all, the word is almost always used against young people—especially women. No one calls Bernie Sanders ambitious, but the man ran for president twice. When he ran in the eighties, Joe Biden was condemned for his ambition; no one said that in 2020. All her life, people have flung this accusation at Hillary Clinton.

It's impossible to succeed as a politician without ambition mushing you onward. The same is true in pro sports, business, the arts, or any other field in which you don't just get hired—you have to continually fight to belong. Actors want to keep getting cast; comedians want to sell out shows; football players want to make an NFL roster; entrepreneurs want to grow their business. In these jobs the only way to keep doing them is to be climbing the *next* mountaintop.

The question isn't whether someone is ambitious. Anyone who campaigns for anything is asking people to give them power. The question is, What are they going to use that power *for*? Prestige and validation and greed? In the hope that maybe their father will finally love them? Or is the goal to do good, to actually help the people who need help? I chose to believe I fit in the latter category, and I really hoped I was right.

I realize, however, that not everyone wants, or needs, to run for the US Senate at age thirty-three. All Stephen wanted, for instance, was to represent Boone County in the Missouri state senate. He would have been happy doing that, taking care of people he cared about. The only place I could find solace was in the thought of the next thing to achieve—I constantly told myself I'd feel better when I hit this quarterly fundraising goal, when I drew even in the polls, when I won the Senate race, and on and on and on.

The press constantly referred to me as "a young man in a hurry." Of course I was. The present was dull and flat and empty. I had to live in the future.

DIANA

In the limited time we got to spend together, Jason would physically be there, but mentally he'd be checked out. I kept trying to snap him out of it, and into what was happening in the current moment, but it was hard for him to get there.

"I'd love to spend some time in the present with you," I'd say.

And his response would be something like "I need your help to do that."

It felt like he had mentally put himself aboard the Nebuchadnez-zar *in the movie* The Matrix—*nothing mattered but the mission. Not the food he ate, the clothes he wore, or his annual doctor or dentist checkup. For him everything in the present seemed to be in gray scale. He wanted to tell me stories from the past or pepper me for reassurances about the future. Never the now.*

Weeks before announcing my Senate run, I'd flown to Washington to meet with Senate minority leader (and aspiring majority leader) Chuck Schumer. The meeting had been set up by my senator and longtime friend Claire McCaskill. Right away, Schumer told me he was already sold on me because he trusted Claire. Then he brought in Senator Jon Tester, who was running the Democratic Senate Campaign Committee (DSCC) at the time. The three of us and Abe started planning, and then I went home to Missouri. In the weeks before my announcement, I received phone calls from just about every Democratic US senator. They were offering help and, I think, curious to see for themselves why Chuck and Claire really thought this thirty-three-year-old kid could somehow win in deep-red Missouri.

Of course I knew that a Senate campaign was a whole lot different from a regular statewide office campaign, but it really came home to me when the staffers from Washington descended. Having the blessing of the DSCC meant *some* money, but there wouldn't be much of that unless I was able to make the race competitive at the end. But when it came to unsolicited advice, I was a rich man. Some of it was helpful, and some of it was what you'd expect from people who aren't from your state.

The DSCC sent a policy person and a communications person to help with announcement week. I did need the assistance with policy: over the previous two years, I had pretty much checked out from the minutiae of national politics because I was laser-focused on state-level issues. To unwind at night, instead of tuning in to CNN, I'd been watching the MLB Network news—in the off-season, no less!—because honestly, I cared a hell of a lot more about where big-time free agents were going to land than what was in the federal highway bill or which of Bush's federal tax cuts were getting extended. Not every politician is up to speed on every issue—and sometimes they don't even understand the issues they themselves oversee (if you've ever watched tech executives testify in front of mystified House committee members, you know what I'm talking about).*

For the next year and a half, I worked like my life depended on it. I was gone four to five days every week, driving across the state for events or flying across the country for fundraisers. I flew commercial; I carried barely any luggage; when I stayed in a place overnight, it was often at a friend's or family member's house, and in the morning I'd throw my white button-down shirt in the dryer for ten minutes with a damp towel to get the wrinkles out, then head to the next fundraiser. I started wearing anti-motion-sickness bracelets so I could make fundraising calls in the back of the car or a taxi. I carried a briefing binder with me everywhere I went. That binder was my whole life. It included all of the talking points I was supposed to memorize when I "had time" and the call sheets I was supposed to

---

* There are some politicians who genuinely study more than most, and—in my experience—most of them are women.

get through in a day. I started calling that binder my "rifle" because I had to carry it with me everywhere.

Half of my time in these cities was spent being ferried around in Ubers—often five or six a day. This meant that five or six times a day (especially if the driver "fit the profile") my brain treated me to a screening—with surround sound and vivid color—of the time I nearly murdered an innocent Afghan man.

It had happened nine years prior, on Camp Eggers in Kabul. The military had retained local nationals to shuttle us from our base to our safe houses, and we were never supposed to ride by ourselves—you couldn't know if the driver had been paid off. But on one occasion, I broke the rule. It was after midnight, I'd been working since five a.m., and I was in a hurry to get back to my bunk and catch a three-hour nap. There was no one else around, so I shrugged and climbed into the shuttle by myself. After the vehicle exited the front gate, the driver suddenly turned and went a different way than usual. Alarm bells went off in my head—this was one of the ways soldiers wound up kidnapped or killed. I began by firmly asking the man where he was going. He didn't respond, so I started yelling at him to turn around, but he still ignored me. By this point, I was envisioning the Taliban snatch crew waiting around the corner to tape my mouth, bag my head, and throw me into a trunk for an appointment with a decapitation video.

Frantic, I put my pistol to the back of his head and screamed at him to stop the vehicle, but he kept going, shouting back at me something I didn't understand. I was thinking, *Do I really need to blow this man's head off and run for it?* I knew I had only seconds to decide. And just as I was steeling myself to do that, I looked up—there was the back of my safe house.

The route had been changed, and no one had told me. I had been threatening to execute a man whose only crime was not speaking English, a language he had no use for until we invaded his country.

I caught my breath and holstered my pistol. "Sorry," I said, because what else was there to say? He didn't understand. To this day I wonder how much trauma I inflicted on that man as he tried to sort out why that crazy American maniac was threatening to kill him.

Time and again, when I was headed to a meeting with a prospective donor, the panic in the Afghan man's screams would come rushing back and wipe everything else from my mind. In the elevator I'd have to ask my staffer, two or three times, to remind me of the name of the person we were about to visit, even though I am blessed with an excellent memory.

And then, if the meeting was going well, those in the room sometimes asked to hear about my deployment, occasionally flat out inviting me to share a story. For a moment, I would wonder what it would be like to tell a donor the story that had been playing on a loop on the way over—the one I thought people actually needed to hear. To sit there and talk about preparing myself to murder an innocent man, and watch them shrink into the expensive upholstery. Maybe they'd cut an even bigger check just to get out of the room. Instead, I told them about the time Sabet brought a local official to a meeting. Don't worry about him, he'd said, "he doesn't speak any English. He's an idiot." Midway through the meeting he pushed it further. "This guy is terribly corrupt, and tried to kill me many times, but thankfully, he is an incompetent." Ten minutes later, an FBI agent who'd accompanied me walked out with the man, who immediately started chatting her up in flawless English. He owned land in Nebraska. That one always got a laugh.

When I wasn't on the road campaigning, I was a telemarketer. I spent my days in that tiny carpeted room at the office, with almost no natural light. I'd have a headset on, and across from me sat my call-time manager, taking notes and cuing up the next call, so that as soon as I finished one, I'd launch into the next. Generally, we were both dialing at once, and if one of us connected, the other would hang up. Often another staffer would be pulled into the call room to dial a third line. A TV played music videos or old Royals games on YouTube. I wasn't allowed movies because they slowed me down.

I developed a lot of little tricks to motivate myself, like covering the walls with encouraging notes people had sent in the mail with their contributions, everyone from US senators to over-the-road truckers who donated ten bucks and said I made them feel hopeful. We set hourly goals and kept track of them on a whiteboard. I had a "happy light," which mimicked natural light, but it was pretty small and ineffective. Diana made a giant collage out of photographs from our first campaign and arranged them in such a way that when you stood back from it, you could see they formed a Royals logo. It hung on the wall in the call room, and I often stood before it and studied each little picture while the phone rang and rang. I was hoping to reconnect with our first campaign's sense of fun and adventure.

But more often than not, my lower back would be in so much pain that I'd do the calls lying on the carpet, on an ice pack, with my feet up on a chair.

It wasn't just my back. I wasn't exercising or sleeping, I was traveling nonstop, and I ate like a raccoon bingeing on trash. Actual sores were opening up on my scalp because I couldn't stop scratching it, and my suits didn't fit right because I was losing weight. I was a wreck. People kept asking Abe, "Is he secretly sick?"

It was like I stopped being human. I was just a vessel for a campaign, one that was challenging my idealism more and more every day. I'd imagined that a run for the US Senate would make me feel like the army had—that sense of participating in something bigger and more virtuous than anything I could do on my own—but it just . . . didn't. On long flights I had to rewatch *The West Wing* just to retain some semblance of optimism.

I had always been close to my parents and my brothers, but now I barely spoke to them, had no idea what was going on in their lives. I was aware that weekly family dinners continued at my parents' house, but I was present once every few months at best. And when I did show up, I was too exhausted to be any fun. My dad had been diagnosed with a degenerative motor-neuron disease, and my mom was battling breast cancer. I felt enormous guilt about not being around more.

As I was pulling into the St. Louis airport for an early morning trip to Washington, I learned that my grandmother had passed away. I could think of nothing else during the day's meetings.

I was trying to navigate the tension between being the best candidate I could be and being any kind of remotely present husband and father; I knew how hard I was failing. But I felt it would be selfish if I micromanaged my schedule to allow myself an occasional break. I didn't bother to ask "Is that the best use of our time?" as new tasks were suggested. I figured I had to say yes to everything.

I was jealous of people—the normal people—who could be happy doing a regular job, raising kids, and diving into hobbies. I didn't understand them, but I envied them all the same. I told myself I wanted to move home to Kansas City with my family and do a job that felt meaningful every day. I wanted to have family dinners each night and coach Little League and drive a pickup truck. When I really let myself fantasize, I told myself that maybe, just maybe, I could join a competitive adult baseball team someday. Not rec league softball, but actual baseball.

But the truth was, I just wanted to be a person who wanted those things. Someone who didn't fear slowing down and could imagine actually feeling fulfilled by life.

Instead, I flew to Los Angeles a few hours after the excruciating experience of putting down our dog Winston. We sat with him on the floor and held him as his eyelids got heavy and he faded away from us. Fourteen years earlier, when Diana and I first moved in together, Winston had joined us. We treated him like a human baby in that obnoxious way that young childless couples pamper their first pets. So when he died, it was like a piece of *us* had died. And yet right after the appointment, I had to go.

As I was deplaning in LA, pulling my rolling suitcase out of the overhead bin for the hundredth time, a thought went through me with the ringing clarity of a struck gong: *What the fuck am I doing?*

I knew my approach wasn't normal, but everyone else seemed lazy. Other politicians didn't have my drive because they hadn't been taught the art of hustle the way my dad taught it to me on the baseball field or the way the army ground it into me. The few politicians who *did* seem to work as hard as I did were the only ones I really respected, but of course, they probably weren't completely failing as a spouse or parent. Did politics simply weed out those who lack the work ethic, the hunger? Every day you're forced to answer the question *How badly do you want it?*

The problem is, campaigning for office does nothing to prepare a candidate for the work of governing. During a campaign at least 90 percent of what you do is raise money. A person's campaign skills are a worthless measure of how well they will legislate. Imagine if army basic training consisted entirely of standing at attention for nine weeks. Sure, you would have proved how badly you wanted the job, but you'd be pretty useless in a fight.

I sound like I'm complaining, but the truth was, I liked campaigning. I was good at it. I enjoyed meeting people and performing—it energized me. When I traveled, pretty much always a member of the finance staff accompanied me, so I wasn't entirely lonesome. I also found myself meeting famous people, and not just politicians. Growing up, the most famous person I'd ever encountered was Bryan Busby, a local meteorologist who visited my school.* Now I'd meet someone like the TV executive and host Andy Cohen, who was from Missouri and held a couple of fundraisers for me. I'd had no idea who he was because he wasn't on MLB Network and I hadn't had time to consume any popular culture since the campaign began. According to Andy, now that I'd met him, it would seem like he was everywhere, and he was right. Dude seemed to be on every screen or magazine cover I saw. Apparently he got a kick out of the fact that I was supposed to be the hip young candidate, but I had the pop culture knowledge of a baby boomer who didn't own a television.

I even found a way to like call time—usually I was talking to pretty interesting people. In general, fundraising activities were measurable, and if I hit my numbers for an hour or a day or a month or a

---

* My great-uncle John, a well-known Broadway composer, was actually the most famous person I knew, but to me he was my uncle. The performers I'd met backstage at his shows, like Liza Minnelli, were just my uncle's friends. Like most kids, I was oblivious to many things.

quarter, the pressure in my brain briefly released. I had hit a goal. The pleasant feeling was fleeting, but it was worth chasing.

Throughout the campaign, I never said no to anything Abe or anyone else asked me to do. I was terrified of disappointing them. I felt deeply indebted to everyone who worked for me; I knew I didn't deserve their dedication. So I never put my foot down and refused.

There were two things I struggled with badly. First, I could not sleep on planes. None of us understood why. I wasn't afraid of flying, and in the army I'd learned how to sleep just about anywhere. The campaign staff tried to train me to take red-eye flights when traveling from the West Coast, but I would show up looking haggard and gaunt; I couldn't so much as close my eyes during a flight. Eventually, we gave up on red-eyes completely.

The second problem was the tracker.

Trackers are young staffers dispatched by a rival campaign (or PAC) to follow an opponent and record their every move and word, hoping to piss them off enough to catch them saying something to use against them. Often, they'll shout questions, hoping to get footage as you refuse to answer. Then they'll package the clip to make it look like you're running away from a member of the press. Trackers have to be tenacious, shameless, and slippery. At a certain level of the game, they're a fact of life for politicians. The Blunt campaign put one on me right away (and we eventually had one on him), which was a particular headache. I had to constantly leave the state to do fundraising, but I couldn't let the tracker catch me because I wasn't supposed to be out of state. So on the one hand I was trying to build recognition, but at the same time, in this circumstance I was trying desperately not to get recognized. It felt like I was doing something dirty or wrong. Once, a man recognized me on the street in New York City and got very excited, telling me he was a big fan—and believe me, being recognized in New York City was a pretty big deal for me! But all I could think to say was "Please don't tweet about this."

I knew that the tracker was coming. I knew that rule one of dealing with the tracker was *Do not let the tracker get to you.* I'd put on a good show of nonchalance for the DSCC comms person who'd tried to train me. I thought I'd be fine.

Whenever the tracker appeared, I'd turn to a staff person and start talking about the Royals. I never covered the same ground twice, prided myself on bringing up something new about the team each time. The idea was to play a game, subversively turning the tracker into a deeply knowledgeable Royals fan. But the truth was, I did find this person distracting, and I wasn't sure why. When the tracker caught me off guard, showing up somewhere unexpectedly, my heart would race and I'd break out in a sweat. At a parade or some other public event, it wasn't much of an issue—I expected them to show. But if the person found me when I didn't want to be found or showed up at an event that was supposed to be kept confidential, it destroyed my concentration. The team would reassure me—the tracker wasn't getting any damaging footage—but it didn't help.

After I'd made a disparaging remark about my opponent, calling him out for "using veterans as props," the Republicans sent a gentleman who supported my opponent to confront me. The man's name was Dewey and I knew him from the state capitol, where he frequently represented the VFW. Dewey and I had a lively conversation, with lots of gesticulating and such. To a passerby, it may have looked as if we were arguing.

Later that day, I found out that a tracker had been hiding in the bushes across the street, with a zoom lens. When I saw the footage posted to YouTube, with the title "Jason Kander yells at veteran," I instantly knew no damage had been done politically; there was nothing scandalous about two veterans having a conversation. But that didn't stop me from feeling I'd been lured into a trap and allowed someone to have me in their sights. After that, I became even more distracted by the possible presence of trackers, though I objectively knew I wasn't saying or doing anything I wouldn't be comfortable seeing on television.

In the nightmares I had every night, if anything—even a minor item like a piece of paper, a water bottle, a toaster—was out of place or unexpected, it meant I or someone I loved was about to be taken.

I understood that a connection existed between my nightmares and my reaction to the tracker, but I decided there was nothing I could do about it. I had no inclination to explain it to people who might think I was nuts. Though I was starting to think I was nuts.

Whenever I spoke at an event, I made sure to shake the hand of everyone in the room, even when a thousand people were there. I appeared to be a tireless practitioner of retail politics. And I was. Connecting with people was genuinely important to me. But that was only part of the story. I didn't feel fully comfortable about mounting the dais and taking my place at the podium until I'd assessed every threat in the room—until I'd shaken its hand and looked it in the eye.

Meanwhile, True and Diana were left alone in a big empty house in Columbia, MO. But when I called home, saying "I love you" wasn't the only priority.

"Did you remember to lock all the doors?" I'd ask. "Are all the alarms on? Can you check?"

I became a tyrant about safety. When Diana would drive back to Kansas City with True to visit our parents, she'd call to let me know she was leaving, and for the next two hours, until she let me know she'd made it there safely, I basically wouldn't breathe.

It's hard to explain this, but I'm going to try because it's important. Think of a moment when something awful *almost* happened to someone you love, a moment that came out of nowhere and occurred too swiftly for you to stop—your child is jumping on the bed and falls off and *just* misses landing on her head; your spouse is fixing something in the garage and you hear a crash and you run in and something has collapsed. And in this scenario, it turns out that they're okay, right? No lasting damage. But when your child has stopped crying and doesn't seem to have a concussion, and you've helped your spouse clean up the broken pieces of the thing that just missed their head, you start to think of what *could* have happened. The broken neck. The shattered skull. Out of nowhere, your life, devastated. For the instant that this feeling soaks into your bones, it's like you're plummeting, choking on uncontrollable dread.

Now imagine having that feeling constantly, regardless of whether someone just fell or something crashed. Imagine feeling as if everything is a threat and every non-event is a lucky escape—a close call. I was certain danger was all around me, and I had to be constantly preventing, thwarting, and controlling it. I wasn't too intense; everyone *else* was tragically naive about just how dangerous the world really was. Worse than that, I was slowly turning Diana into a person who saw all the same menace in the world that I did.

In 2014 a neo-Nazi member of the Ku Klux Klan drove to a Jewish retirement home not far from where Diana's parents lived in the Kansas City suburbs and killed a woman in a parking lot. He did this immediately after murdering a man and his grandson at the Jewish Community Center down the street. The man and his grandson were the father and son of Diana's friend Mindy Corporon, who just a few months earlier had created a special gift for us: a blanket she'd embroidered with True's name.

Meanwhile, the capitol police in Jefferson City informed us that they had been monitoring the murderer because he had interacted multiple times with the secretary of state's office. He could just as easily have chosen me as the target of his anti-Semitism, they concluded. That's when Abe and I decided to hire Chuck Walker, a retired police officer, to serve as my bodyguard. While not an absolute guarantee against danger, Chuck's presence did put me more at ease about my own safety as I traveled the state.

But Chuck wasn't at my home at night. My family was the most visible Jewish family in the state of Missouri—and anti-Semitism is baked into our family story. Alone with True in our empty house in Columbia, Diana became convinced we were the next target. Moreover, every time I went out to campaign or traveled in my capacity as secretary of state, my office sent out a press advisory. That's what every politician does—but to Diana, it meant I had just broadcasted to the entire state that my home was unguarded.

It didn't help that when I *was* home, I was frequently an asshole. Out on the campaign trail, I was charisma itself: chill, funny, sincere with voters. With staff, with donors, I had to be Jason Kander, the Easiest Person in the World to Get Along With. Absolutely no one suspected what was going on inside me. I never got mad at anyone—except Diana.

"I wish I could get the version of you that everyone else gets," she'd say. But I couldn't give it to her: she was the only person in the world who was getting the real me, the one who—just beneath the surface—was angry and terrified all the time. And she had to bear it all by herself. When I was home, I was rarely emotionally *there*—I just felt nothing. When my feelings did rear up, it felt like they were trying to swallow me whole. Ever since True was born, I'd struggled to be emotionally present, and mostly I'd failed. I did witness some important moments with True, such as his potty training, yet I wasn't completely there. Once, True was so proud of himself for peeing in the toilet that, while celebrating, he pooped on the floor. I objectively understood that this was hilarious, and I surmised that a human ought to feel something! I could see it, but I couldn't experience it. Instead, I was observing everything from a distance; and I judged that guy standing there, holding his laughing toddler. That guy was such a garbage human being, he couldn't love his son properly. More evidence of my inadequacy, more fuel for my shame.

I'd been home from Afghanistan for eight years. It seemed wrong to continue to blame it for the problems I had; they had to be my fault. Slowly, I began to accept that I didn't feel happiness anymore. It just wasn't for me. All I was good for was working and winning. There was only one way I ever earned a break from the anger and shame: work myself to exhaustion, just like I had done four years earlier. When I was fully drained, when I hadn't slept and my back hurt and I was completely hollowed out and still going—that was the only time I felt like I was worth a shit. So day in, day out, I drove myself to that point.

I felt I was destined—or more like resigned—to a short life of consequence. One that mattered to others but was never really experienced by me. Because I didn't sleep, ate poorly, and didn't exercise, I was frequently sick and pretty much always in pain. To me, this meant I probably wouldn't live very long. I just couldn't picture myself making it into my nineties like my very active grandfather or great-uncle; the way I treated my body, I didn't see how it could hold up.

Often, on the campaign trail I met veterans who would tell me they hadn't felt right since coming home, that their marriage had fallen apart, or that they'd had trouble holding a job because nothing in civilian life felt meaningful. Over and over, I sat and listened and hugged them and counseled them on how to get help. In many ways these conversations kept me going—I wanted to help these fellow soldiers—and I could see clearly that they needed it, and there was no shame in it. I told them as much. But I didn't think I was like them. They had earned compassion and I hadn't. Even if my problems had once been service-related, that was irrelevant now. I was in politics. I couldn't afford to be like them.

Meanwhile, my aspirations had now run headlong into Diana's. Her success as an entrepreneur had inspired her to share what she'd learned, so she wrote a book on the subject,* which catapulted her into a new career as an innovation consultant to major corporations and a paid speaker on the conference circuit. She loved this work, but it required traveling, and fitting her travel around mine was nearly impossible. She wound up turning down a lot of projects she'd have loved to do. She made the best of it by accepting a position as an associate teaching professor in the MBA program at the University of Missouri. She enjoyed it, but there was no doubt she had put a restrictor plate on her consulting and speaking work.

It didn't help that my campaign staff—as much as they loved her—viewed her career as a liability. The DSCC apparently had some sort of Google alert set for Diana's website. We learned this when she posted—like every other consultant in the world—a list

---

* When she decided to write a business novel, I was skeptical, given that there wasn't a section for that genre in bookstores. Today, that first book she wrote, *All In Startup*, is taught in over a hundred universities and has been translated into several languages.

of her past clients. The DSCC called my communications director within minutes, apoplectic: "Why is she doing this to us?" My communications director brought this concern to me, and I had to call Diana and ask her to remove the list. It was unfair, and I felt terrible about it, but when she protested, I said, "This is doing the other side's opposition research for them." Reluctantly, she took the list of clients off her website.

As if that weren't enough, the other side ran ads claiming that she was a lobbyist, which she very much was not and never had been. People would recognize her in the grocery store and—instead of asking about her bestselling book or what kind of consulting she did—make polite conversation by asking who she lobbied for.

Meanwhile, out in Columbia, we were two and a half hours from family and constantly between nannies; I was rarely home. Diana sometimes lost entire weeks of work to raise our son, practically without me. When she turned to me, looking for a safe place to vent, I would get defensive, which sparked an argument.

Diana

I had a recurring nightmare about being responsible for the end of Jason's career. We'd be on one of those ferries that goes out to the Statue of Liberty. Around us, people were crowded in like sardines. This attractive woman standing next to us would lean in and sniff Jason. And feeling a wave of emotion and no way to control it, I would bite her on the cheek. She'd look at me, astonished, cover her cheek with her hand, and run off to show the wound to everyone, and that would be it—everyone knew Jason Kander's wife was a feral maniac. That was the end of his career.

I was so worried about getting in the way that I'd try to be as careful as possible. I never mentioned to anyone the struggles Jason was going through. I made every effort to avoid doing anything professionally that might have the slightest appearance of impropriety.

I tried to be the best supportive spouse that I could be, but there was no denying that my anxiety and anger were growing. I felt isolated and alone, like I had no footing. I felt like an outsider looking at all of the things Team Kander was doing. Not only did I no longer feel like I was really on the team, I felt like the team saw me as a liability, or even worse, a rival. That was very disheartening.

I was also confused by interactions with Jason. At night, we'd be talking about something inconsequential like the dishes, and suddenly we'd be arguing. I wouldn't understand why, or even what we were arguing about. We're both experienced debaters, but he would swerve from topic to topic so fast, I couldn't follow him. All I knew was that he was angry and I was angry too. The only way out seemed to be a complete shutdown, a withdrawal, and a hope that things would be better in the morning.

I really wanted to get more exercise. Not to get into shape, but to shut down my brain. I desperately craved the calming effect of exhaustion that I had discovered when I did mixed martial arts back when Jason was in Afghanistan. But it was hard to figure out the logistics with a baby at home and Jason on the road. Toward the end of summer, I cleared out the dining room (we never used it anyway) and put in a rowing machine that I borrowed from Stephen. Every night I rowed for a good thirty to forty-five minutes—no music, no headphones, so I could hear the baby or any other sound. Rowing and rowing in silence. It was my only reprieve.

———

Of course, the thing that dramatically altered my campaign came back to my army training. In the field, I had never once needed to strip and rebuild my rifle blindfolded. But I could do it—I'd cleaned my rifle in the dark plenty of times. So we decided to make a video of me as I did so, while chatting amiably about why I supported gun control. When we put that video on television, it *erupted*.

That ad turned loose a firehose of money and national attention—I no longer had to do my How I Can Win spiel for donors. I was now "the guy from the gun ad," and that opened a lot of doors—or more accurately, a lot of checkbooks.* As it became clear that Donald Trump was going to be summarily blown out by Hillary Clinton, attention turned to taking back the US House and Senate. Now when I showed up to events, actual crowds were waiting to hear me speak. For the first time people started asking me for my autograph—and for selfies too. We had to schedule a half hour at the end of every event for a selfie line, something unheard of in Missouri. And I bathed in the attention. The more of it I got, the more momentum I felt, and the better I performed. It energized every aspect of the campaign.

But every time someone introduced me at an event, the lore of my wartime service got inflated a little more. I'd politely correct any overstatement as I began my speech. ("It's nice of you to say 'highly decorated,' but honestly, I was 'lightly decorated' at best." "I appreciate the kind words, but I do want to clarify that I served only one tour, not multiple tours like a lot of my friends.") On top of that, I had just become famous for an ad featuring my deft handling of a weapon I'd never fired in combat. All of this was grist for my guilt mill. I felt I was being given an honor I did not deserve, and I worked even harder in the home stretch to dull the shame. Every time I wanted to slow down, every time I considered feeling sorry for myself, I'd remember this: *T.J. can't fucking see and Kevin is dead.*

The watch party for my Senate race felt like a wedding—everyone I knew and loved showed up, ready to celebrate. That entire day, I felt great. We were winning. Our final polls had me in the lead by a few points, and national publications were starting to project Missouri as a flip; some said Republicans in Washington had written it off as a loss. When the polls closed and the numbers started to tighten, all I could feel, once again, was terror: I was going to die. In my mind, I saw a shovel.

---

* Though people tend to remember that ad as making the race competitive, that's not actually the case. We ran it for a short time only, and we had pulled slightly ahead in the polls before it ran. It got a lot of attention, which helped, but it mostly helped us outside the state.

I was in a large, deserted building in Kabul, meeting with two men. They were dressed in Western suits, and one was loading wood into a furnace. Salam and I had no backup, no one securing the door two floors below, no idea who or what was in the myriad other rooms on this floor alone. It was evening, and the sun had set since we'd arrived. Being out after dark was never smart.

As I entered the building, I'd seen a shovel in what used to be an elevator shaft on the first floor. Why did these two need a shovel? And why did they need an entire building to house just one warm, ornately decorated office? The men were members of the Afghan government tasked with curbing drug trafficking. They were also drug traffickers. We were alone, and I was completely vulnerable. *If they never find you, what do they say to your wife? How will Diana tell my parents? Will she call, or will she drive over? My dad will probably be the one to tell my brothers and my grandparents, I guess? He'll probably stay strong in front of others. That's what Dad does.*

But I wasn't in Kabul now. I was in Kansas City, but like four years earlier, I felt like I was taking my last breaths. Only this time, no relief came. The *New York Times'* needle wobbled. Stephen lost his state senate race in Columbia; we talked on the phone, and he sounded dead. Hillary lost Michigan. Then Pennsylvania. Then the presidency. For a few minutes, I was still alive, and then I wasn't. A phone was put into my hand. On the other end was the enemy. "Congratulations," I said, admitting my defeat. I could not believe the ceiling wasn't collapsing. This time there would be no midnight rescue.

The next morning the fact that I woke up alive seemed almost . . . insulting. That I could feel *this* bad, this close to death, and be expected to live through it felt like a sadistic joke. As I kept saying to Diana while lying in bed and staring up at the ceiling, I felt like I was dying. That was the only way my brain could process this. Imagine feeling like you are about to die. The airplane is plummeting, the wind screaming in your ears; the water is filling your nostrils; whatever horror comes into your head when you imagine that moment, *that* is what I felt, only it wasn't like a thought in my imagination. It was real. I'd let the whole country down.

Somehow I found myself in a booth at a breakfast place. Food was put in front of me, as if I had any right to eat. Eating seemed insane, trivial. Abe and Diana were there, saying things. My phone rang. It was Jonathan Martin from the *New York Times*. Out of habit, I asked Abe whether I should take it, and the look on his face said, *Now that you died, who cares?*

I went outside. I said, "It is what it is." I pretended to be Jason Kander—one more for the road, I thought—and gave him a quote.

A woman saw me and stopped.

Not now. Please don't.

"I just have to tell you," she said. She clearly hadn't slept any more than I had. "My daughters look up to you *so* much. They drove home from college because they wanted to vote for you in person."

She began to cry and wrapped her arms around me. I felt myself start to cry, and I fought to hold it in, because once it started, I knew it wouldn't stop. Shame slid into my stomach like a white-hot bayonet. I'd let down this woman and her daughters. I had let everyone down.

7

THE COOL KIDS' TABLE

Barack and Michelle Obama's office in Washington, DC, isn't grandiose or flashy. Like the Obamas, it is classy, appealing, and understated, which is to say that other than the Secret Service presence, it has the feel of a successful boutique law firm where every employee is either a founding partner or a founding partner's adult child. It was January 2018 when I entered the small lobby outside the former president's private office and began signing in on a tablet mounted to the wall. A guy about my age came walking down the hall toward me. As we shook hands, he could see that I was still searching my mental files for a name to put to his face.

"Pleasure to meet you, Jason. We have a bunch of mutual friends. I'm Ben Rhodes."

Ben had been deputy national security advisor in the White House. He was right, we did have a bunch of mutual friends, almost all of whom I had gotten to know only in the past year.

"We're excited you're here."

"Thanks," I said. "I'm obviously excited too, though I can't say I have any idea how I ended up here. Are you one of the people in this meeting?"

Ben gave me a smile that said he knew something I didn't. "Nope, but I think it'll be a good meeting. I know the boss has been looking forward to it," he said, before excusing himself to go to an appointment.

I sat down in the waiting area, wearing suit pants for the first time in months. I hadn't even worn suits pants on the campaign trail toward the end—rocking jeans and cowboy boots had been my micro-rebellion against the constraints of the race. But when Barack Obama summons you, you put on a suit.

Paulette Aniskoff, who led President Obama's post-presidency political operation and had been my main point of contact so far, sat down next to me. "So, how ya doing?"

"I mean, pretty fantastic," I said.

"Well that's good, why fantastic?"

"I mean . . . I lost," I said. "What am I even doing here?"

She laughed and said, "I think you'll enjoy the meeting."

This was weird. I had tried not to read too much into the invitation. Obama was settling into his post-presidency. I figured I'd get about fifteen minutes with him and a staffer or two, and that I'd be afforded the opportunity to make one ask. I was prepared to request that he appear as a headliner for a fundraiser for Let America Vote, the anti-voter-suppression project Abe and I had gotten up and running over the past year. If that was a no-go, my fallback request was for him to sign on to one of our fundraising emails.

Finally, the front door, the one I'd stepped through a few minutes earlier, swung open, and Obama strode in, dressed in dark jeans and a black sweater. "Sorry I'm late," he said, though he was on time. I had arrived early. Instead of a handshake, he wrapped me up in a bro-hug and invited me in. The wood-paneled office was spacious and neat, with a smattering of keepsakes and family photos behind the desk, upon which sat only a few pieces of paper, a pen, and a closed laptop. He gestured for me to take a seat on an elegant gray couch. Then he closed the door and sat down in an armchair to my right. To my surprise, no aides were present.

He began by teasing me for wearing a suit, and I joked that I was glad he had noticed, since it wasn't something I did lightly. A few minutes into an in-depth conversation about the origin of neckties and a shared mutual frustration with their staying power in men's fashion, it became clear that he was in no rush to wrap this up.

I'd never dreamed, when I woke up that morning, that I was in for a long, friendly head-to-head with the man I admired most in the world. But then, fourteen months before, it had seemed impossible that he would have reason to remember I existed at all.

I'd spent the afternoon of November 9, 2016, in a therapist's office. The night before, as the scale of the disaster became apparent, Diana had snuck away from the Senate campaign "victory party" to book an emergency appointment. After the number of times I'd asked her whether she and True would be better off without me, she knew how dangerous this loss could be.

I know that a lot of people cried that morning, and for days afterward. But at first, even when the therapist started asking me questions, I held everything in. I'd gotten good at that. Then he tried some semi-hypnotic thing, asking me to close my eyes and visualize myself as a child, playing in the front yard. "What," he asked me, "would you want that child to know?"

I saw a little redheaded kid tossing a baseball up in the yard and catching it in his glove. "The world will ask a lot more of you than you expect," I said. "And you won't always succeed, but you'll keep trying . . . and that means you'll miss out on a lot."

That, at least, unclamped my tear ducts. It felt a little cheap yet also good to let myself cry silently for a moment. Then the therapist called Diana into the office and gave her a sort of troubleshooter's guide to her malfunctioning husband.

"He's depressed," he explained, and I wasn't going to argue with him: I felt like hammered shit. But my next thought was *So you lost an election? Big deal. What right do you have to be depressed? You came home from Afghanistan alive. Suck it up, asshole.*

That evening, Diana said, "You know, people still want to hear from you."

"No, they don't."

I wasn't just wallowing. I meant it. Why should anyone care what I had to say anymore? Elections aren't baseball seasons; you don't just chill all winter, then come back in the spring. I had seen plenty of flash-in-the-pan national-sensation candidates lose, and I had never seen any of them do anything but fade into irrelevancy. They might as well be dead. Remember Howard Dean? His political career was dead, so he was dead. I didn't care if he was living a happy life in Vermont. He died.

All of the attention, all of the media, all of the spotlights and speeches—all of the *meaning*—it was gone. Overnight, the phone stopped ringing.

Sitting there like a lump of useless human matter taking up space on our couch, I had been aware—even before Diana's prompting—that a candidate was expected to send a farewell missive to supporters, but I dreaded writing it. My mind went back to the letter I'd written to Diana ten years earlier while hiding in the bathroom of our house in Waldo and crying. I could still remember the opening line: "If you're reading this, I'm sorry."

There's no army regulation requiring you to write an "in case I get killed" letter to your next of kin, but it was generally regarded as the responsible thing to do. In my case, I put it off until the last minute, probably because I resisted the idea of inhabiting the headspace of my dead future self. Now I remembered the experience as overwhelmingly emotional, and emotions were not something I was currently in the mood for.

But worst of all, I now felt like that dead future me. And as soon as that self-pity materialized in my mind, I dismissed it, spit on it, and took it as evidence of my own worthlessness. I was disgusted at myself for hurting so bad over something like losing an election when I had friends who were still serving overseas.

Diana

I knew I had a moment here. A moment in which the outside world had forced us to stop and regroup. This was an opportunity to get help. We could see a couple's counselor, maybe look into anger management, or just do some yoga. At the very least, an annual checkup

Diana and I fell in love in high school and got married at twenty-two. We called ourselves Team Kander because our goal was to change the world together.

All photos are courtesy of the author unless otherwise noted.

Each time my translator Salam and I ventured off the base to collect information about corruption or espionage, we risked walking into a trap. Salam's knowledge made me a better intelligence officer and his savvy helped keep us safe.

My job in Afghanistan sometimes called for me to wear street clothes instead of a uniform, which made me feel like a cowboy. Almost eleven years later, a clinical social worker at the VA would finally convince me that being practically alone for hours at a time in the most dangerous place on earth, with no one knowing where I was and no one backing me up as I met secretly with people who might want to kill me, counted as trauma.

I experienced nightmares and felt a constant, simmering anger during my first campaign for the state legislature. Here I am, fifteen pounds underweight and barely holding my emotions in check during my victory speech. *Eric Bowers*

Marine combat veteran Stephen Webber (left) and I arrived in the statehouse in 2009. We bonded instantly over shared experience, righteous indignation, and dark humor. Years later, Stephen played a crucial role in convincing me to get help at the VA. *August Kryger /* Columbia Daily Tribune

While serving in the legislature, I continued to serve in the army, including three years as a platoon trainer for the Army Officer Candidate School. Though my army service was only part-time, I saw myself primarily as a soldier who happened to have other jobs too.

I loved the army. Unlike civilian life, military life made sense to me. After my request to return to Afghanistan was denied, I made the difficult decision not to serve beyond my commitment. Leaving the army felt like losing the only part of myself that I actually liked.

Abe Rakov ran my campaign for secretary of state in 2012, and over the decade that followed, he ran pretty much everything I did in politics. A great friend, Abe always saved me a seat facing the doors.

When election night 2012 ended in victory, I didn't feel joy. Instead, I felt relief from the sense of impending death I'd experienced all evening. Years later, in therapy at the VA, I would learn about the connection between my trauma and my need to feel in control.

After our son, True, was born, I struggled to be emotionally present with him and with Diana, a situation that fueled my sense of self-loathing. Meanwhile, my paranoia and hypervigilance were worsening, and I became convinced that my entire family was in constant danger.

By the time I ran for the US Senate in 2016, I was on a quest for personal redemption because I believed I had failed to "do enough" for my country. *Suzy Smith*

After the Senate race, Abe and I started Let America Vote, and I found purpose, along with distraction from my symptoms, in the crucial fight against voter suppression. Let America Vote became a real force in states across the nation, including Iowa, where this picture was taken. *Suzy Smith*

2020 HOPEFULS OBAMA HAS MET WITH:

SANDERS	WARREN	BIDEN	PATRICK	BOOKER
LANDRIEU	GARCETTI	BUTTIGIEG	KANDER	HOLDER

#INSIDEPOLITICS

LIVE

OBAMA HOLDING SECRET CHATS WITH 2020 CONTENDERS CNN

In early 2018, President Barack Obama summoned me for a private one-on-one meeting. When the press uncovered this list, a lot of major political donors began to take my potential candidacy seriously.

In 2017 and 2018, I boarded more than three hundred airplanes, gave political speeches in all but four states, and rarely saw my family. I was now emotionally numb almost all the time. The only things that jump-started my endorphins and allowed me to feel anything at all were standing ovations and requests for selfies. *Michael Martin / TwoSeam Films*

On stage with Diana, True, and New Hampshire Democrats after I all but announced my presidential candidacy during a speech carried live on national television. It was everything I'd ever wanted professionally, yet I kept feeling worse. Within a few months I developed suicidal thoughts, prompting me to drop out of public life and begin therapy for PTSD. *Michael Martin / TwoSeam Films*

I grew a beard so that people would be less likely to recognize me in public. For several months, I attended weekly sessions at the VA, did my therapy homework, and gradually regained the ability to be emotionally present with Diana and True.

Diana developed secondary PTSD from living with me all those years, and she got therapy as well. We also got Talia, who wasn't a PTSD service dog but rather a dog with PTSD. Diana had to train her to relax and cuddle.

True showing *Angry Birds* to my grandfather and my dad. My therapist at the VA taught me how to "feel the feelings," and gradually the numbness and other symptoms faded. That gave me the chance to spend meaningful time with my grandfather before he passed away a couple of months after I started getting help.

When I needed help navigating the VA process, I turned to Veterans Community Project. Today I'm leading VCP's national expansion alongside its cofounders.
Veterans Community Project

All those years of neglecting my mental health left me in constant physical pain. Now fitness is central to my mental health regimen.

My career used to be the only thing that quieted the chaos in my mind, so I put it before everything else. Today coaching True's Little League team is one of my highest priorities.

I make my own happiness a priority now too, which is what led me to start playing competitive baseball again. *Samuel Ashworth*

Diana refers to the time since we got treatment for PTSD as our second marriage. It makes me so happy to see her happy again. *Sam Meers*

Team Kander: Diana, True, Bella, and Jason.

was in order for both of us. I knew we needed to take time for ourselves. To find a way to move forward together, like the team we were when we started. I desperately wanted to be back on that team. I believed in Jason and what he stood for, but something big was missing from our relationship.

But I didn't take the opportunity. He was hurting so much. It seemed selfish and pointless to bring up any of this. So I tried to make him feel better in the only way I knew how, by encouraging him to get right back to the work and the lifestyle that were driving us apart.

It was like we'd been swimming in the ocean, and suddenly Jason was sucked under by a rip current. In shock and terror, I did the only thing I knew how to do: I plunged in right after him, swimming as hard as I could to save him. But you can't outswim a riptide. There is a reason why lifeguards—the strongest swimmers on the beach— carry flotation devices. I didn't know that when a person's drowning, you don't jump straight into the water unless you want to be swept away with them. It's better to throw them a lifeline, give them something to grab onto and float. All too late I realized that instead of saving him, I'd joined him in drowning.

———

"You inspired a lot of people," Diana insisted, "and I can just sense it. You still have a role to play." I certainly wanted her to be right. And since she was the only person I knew in politics who had predicted a Trump victory from the beginning (I, meanwhile, was telling people that if Trump was the nominee, we were gonna win *Texas*), I was pretty sure her political instincts were currently sharper than my own.

"I guess I have to write one of those 'Thanks for supporting me' letters, huh?"

She shook her head. "No, you shouldn't write a regular letter. I know you don't feel good or helpful right now, but neither does anyone, and you have the power to make people feel better. You can make them feel helpful."

That night, I sat down and wrote a letter to my supporters.

"I'm going to be fine," I wrote, and in that context, I meant it. I wasn't going to lose my health insurance or be deported from the only country I'd ever known, but I also knew I wasn't going to be fine. The only thing I thought might make me feel alive was continuing—in some fashion—to matter, and the only way to do that was to fight. I didn't wholly believe the words. I was just slipping back into the character of "Jason Kander." However destroyed I was, at least I could still move that Jason around; I could still make the machine work.

> *If you were a part of this campaign in even the smallest way, you might feel like stepping away from it all to lick your wounds. Maybe you think you're done with volunteering or donating or even believing in anything changing. Well, you won't get a pass from me. Staying engaged has become more important than ever.*
> *Take some time off . . .*
> *Okay, was that enough time?*
> *We have work to do. You in?*

We sent it to the campaign email list, posted it to social media, and went to bed. I had my usual night terrors, but I did something I hadn't done in a long time: I slept in.

The next morning I woke up at 11 a.m. to discover that my phone was practically bleeding with more messages than I could possibly read. It wasn't just friends and well-wishers. A lot of people—important people in the Democratic Party—wanted to talk to me about where the hell we should go from here. My Twitter following had more than tripled. It was like emerging from a bunker after nuclear annihilation to discover that everything you ever knew had been incinerated, while also finding weird comfort in the fact that the handful of stunned survivors had elected you their leader.

The letter had gone viral. Diana had been right, as usual.

The months after the 2016 election were dark days for Democrats. The entire party felt rudderless, adrift, like we were on a ship in search of a new crew and a new captain. A couple of weeks after the election, the Senate Democratic Caucus asked me to come speak about what I'd learned and how I'd been able to garner so many crossover votes. A lot of senators—including some of my fellow 2016 candidates who won—spoke in that meeting. I looked around and thought, *I may have lost my campaign, but I would have done well here.* After the meeting concluded, a line of staff formed for selfies with me, the guy who *wasn't* a senator.

After that, Elizabeth Warren told me something that helped revive me: "It doesn't matter whether you won or lost. You're part of this party now. You're one of its leaders." She was telling me, *No matter what you choose to do, you're one of us now.* It meant a lot to hear her say it, and I tried my best to believe her.

A month later I got a call from a guy named Matt Sinovic, who had been the catcher on my high school baseball team. He was in politics now, and he had a speaking slot open at an event that he needed to fill. But it wasn't just any event. Matt was the director of Progress Iowa, a well-regarded group of activists.

I was excited. An invite to speak in Iowa, even if I was just filling a slot, felt like being called up for a tryout with the Royals. Here was a shot at some kind of redemption. Abe and I took the themes we'd played with in my letter and mashed them together with the victory speech I never gave. The result framed a way forward for Democrats, a way to get up off the mat and take the country back. I used my experience as an example: While Hillary Clinton had been blown out by 19 percent in Missouri, I'd lost by only 2.8 percent—in other words, 16 percent of Missourians voted for Trump, and then voted for me. If I could do that in Missouri, then Democrats could do it anywhere. All we had to do was something simple and radical: tell voters what we actually believed, and fight for it.

The speech didn't bring the house down—this was a "holiday party" where no one felt any cheer—but the reception was warm and appreciative. It was very Iowa. Giving the speech brought me back from the dead, somewhat. When I stepped off the stage, I felt like when you're playing *Mario Kart* and you go flying off the side of a road, and a little turtle on a cloud fishes you out of the abyss and plops you back onto the track. I was still in the game.

Then another call came in. This time it was Donna Brazile, outgoing chair of the Democratic National Committee. We'd never met, but she explained that the DNC meeting was coming up. "There's going to be a big fight for party chair," she said, "but everyone is depressed. We need to show the press that the party can still engage young people; we need new blood. I know it's short notice, but can you come to Atlanta next week and give the keynote?"

Of course I could. It wasn't like I had anything else to do.

Backstage at the DNC, I was feeling downright resurrected, especially when Abe and I ran into a (very much still alive) Howard Dean. Stephen was there too, having just been elected chair of the Missouri Democratic Party. When I got to the podium, I saw Stephen positioned in the front row to offer moral support. I gave a tightened version of the speech I'd tried in Iowa, and this time, I crushed it.

Now the invitations to speak began to roll in. This was new. I'd been traveling the country for two years, but always to donor events, where I'd have to justify my existence over and over again and then ask for money. Now when I traveled, it was to speak at colleges or big gala dinners at state party conventions, and I was almost invariably the main speaker. I felt like a rock star on tour. It wasn't just fun, it was addictive. There were no trackers, no polls, just big crowds of people who sometimes knew who I was. And the speeches weren't asking for votes, either. Now I was speaking to the Democratic Party itself, urging it to follow a new path forward. It didn't feel self-indulgent; it felt like leadership. I loved it. It offered me what I couldn't offer myself: a fleeting sense that I might be worth a damn.

I'd become the guy who won by losing. If you've ever seen the Christopher Guest movie *For Your Consideration*, about actors who accidentally stumble into Oscar buzz while making a mediocre movie, you have a pretty good idea of what political buzz feels like. "Buzz" is a word for the noise caused by random vibrations within an echo chamber. It's not real, but it has real implications: CNN gave me a one-year contributor contract and put me on TV whenever I had an opinion I wanted to share, my Twitter following ballooned, and suddenly I had a commanding online platform. I launched my podcast, *Majority 54*, which unseated Oprah's to debut at number one in the country. Soon the *Washington Post* and *Politico* were speculating about whether I was going to run for president. And to tell the truth, I *was* thinking about it. I'd thought about it for the first time on election night, during the one-hour time window after the presidential race had been called but before my lead faded. In truth, I'd been thinking about it since I was fifteen. But also, the whole idea was delusional: a mid-tier state official, barely thirty-five years old, running for president of the United States? Why not just challenge Lebron James to a little one-on-one while I was at it?

One day I quietly asked Josh Earnest, Obama's press secretary, who'd become a good friend, "Is this insane?"

"Jason, Donald Trump is the president," he said. "There's no such thing as insane anymore. Frankly, if you weren't already thinking about it, I'd be encouraging you to start."

So I did. And more important, so did Abe.

About six months earlier, Abe and I had launched Let America Vote. It was a continuation of the fight for voting rights we'd begun in Missouri, but instead of fighting in court like most organizations in the voting rights world, we identified a different mission: creating political consequences for voter suppression. No one had tried to win the political argument over voting rights, but we felt that it was time to try.

Now that this buzz was starting to build, Let America Vote was being sucked into it. We never lost sight of our original mission, but we definitely became more strategic about where we focused our efforts.

Abe opened a Let America Vote office in Iowa, moved there, and began to lay the groundwork for my presidential run. He got a two-bedroom apartment so I'd have somewhere to crash when I visited. In fact, Let America Vote opened offices in three of the first four states on the presidential primary calendar (doing this in all four would have been too obvious), and our top people—like Ben Tyson, Suzy Smith, Austin Laufersweiler, and Brendan Summers—up and moved their entire lives to those states to oversee operations.

The fact that Let America Vote was enormously effective in its voting rights mission factored into all of this buzz as well. We recruited thousands of volunteers, knocked on hundreds of thousands of doors in targeted districts across the country, and unseated dozens of vote-suppressing state legislators. Let America Vote even helped flip two state legislatures from red to blue.

By the time the 2018 midterm cycle began to spin up, we had more organizers on the ground in New Hampshire than either the Democratic or Republican state parties. Abe and I had built an army there, and as result, I had a statewide profile. As one local political reporter told Abe, "I don't know how you did it, but it feels like every Democrat in New Hampshire is trying to get Jason Kander's endorsement."

With every step forward, the idea that I could actually win the whole damn thing seemed a little less crazy. Meanwhile, with Donald Trump in the White House, the world seemed to be growing more precarious by the day, and the stakes appeared much higher than they did two years earlier, during my Senate race. One day it seemed the president was about to start a nuclear war, and the next he'd be defending white supremacist violence. Meanwhile, nearly every day he attacked people's faith in American democracy.

I was beginning to truly believe I might be the best person to unseat the worst person ever to occupy the White House, but because I still thought of myself as a long shot, I felt obligated to throw everything I had at this pre-campaign campaign. This forced a bigger wedge than ever between politician Jason and husband/father/brother/son Jason.

Yet I never even considered that I had a choice. From my perspective, the stakes were too high to entertain any other course of action.

• • •

One night we held a Let America Vote fundraiser at a bar in Brooklyn, NY. We were in a cozy back room the size of a hotel gym, and Kansas City's own Jason Sudeikis, who had been one of *Saturday Night Live*'s leading performers, was the event's host. I stood off to the side and listened as he introduced me. He didn't just talk about the work Let America Vote was doing or the dangers of voter suppression; he also told how we'd become friends. I'd stepped into a whole new reality.

In early 2017, just a couple of months after my post-election letter had gone viral, Jason Sudeikis filled out the contact form on my website. "I don't know if Mr. Kander or perhaps an intern will see this note," he wrote, "but I'd love to buy him a beer and shake his hand sometime." This was after the Senate campaign but before we started Let America Vote, meaning I had no campaign staff—the only person reading those contact forms was me. I wrote back immediately.

We began texting back and forth for a few weeks. Immediately, Diana decided I was being catfished. He and I had planned to meet up in Los Angeles the night before my first appearance on *Real Time with Bill Maher*. Diana came with me on this trip, though she rarely did. Jason had sent me directions for where to go, and during the ride from the airport, Diana kept insisting: "This is a scam. Get your credit card ready."

But there, waiting for us, was the real Jason Sudeikis. He turned out to be brilliant and kind, and we hung out happily for hours, bonding swiftly over our love for Kansas City.

Afterward, on the way back to the hotel, Diana grudgingly conceded. "I mean, that guy *looked* a lot like Jason Sudeikis."

Now, back in Brooklyn, I moved through the small crowd, shaking hands, and I started to realize that half of those hands belonged to people who lived in my TV. Movie stars, comedians, even some of my favorite ESPN hosts. As excited as I was to meet them, they seemed equally excited to meet *me*.

I'd been invited to the cool kids' table, and it made me feel like a cool kid. I'd been close with politically famous people, but this was a new order of magnitude. Now I was regularly texting and talking to people like Judd Apatow and Lin-Manuel Miranda. I went out to dinner with Jimmy Kimmel and Molly McNearney. Chelsea Handler held a Let America Vote fundraiser at her house, where I spoke to an audience of faces I'd seen only in *People* magazine.

After another Los Angeles fundraiser for Let America Vote, Abe and I went out to dinner with Bradley Whitford and Janel Moloney, who had played Josh and Donna on *The West Wing*. They'd been at the fundraiser, along with two of the show's writers, Eli Attie and Kevin Falls, and as we ate, the conversation turned to 2020.

"What we need," said Eli, "is a young vet from the Midwest who's proved that he can win Trump voters."

I sort of laughed. But he was serious, and the next thing I knew, Brad and Janel were actively trying to talk me into it. Here I was, sitting across from Josh Lyman and Donna Moss, and they were arguing that I should run for president.

It was surreal. I had to step out of the restaurant and call Diana to let her know how crazy things had gotten. But what I couldn't say, not even to her, was that the moment I heard her voice, I suddenly felt sickened: I was having a blast out in LA, hanging out with Hollywood stars, but I'd left her and True all alone at home. At that moment I loathed myself for how much I was enjoying the moment. I wanted to be president, and I wanted to be the type of man who didn't *need* to be president, but every step I took toward the first goal put the second farther in the rearview.

The higher my star rose, the crueler I was to myself. As soon as someone told me how much I inspired them, I'd mentally rebut their statement, thinking what a piece of shit I was for leaving my wife and son behind so I could travel. I'd abandoned them, just like I'd left Afghanistan and didn't go back.

True was four years old now. Once, when someone asked him what my job was, he responded, "Daddy flies on airplanes." I wasn't even the guy who piloted airplanes, like my dad did. I was just a guy who sat in a seat on a plane and FaceTimed True when I landed.

Growing up, my dad traveled for work, but I can count on one hand the number of my games he and Mom missed—and I played sports year-round. My dad was my baseball coach and his dad, Pop, had been his baseball coach. I dreamed of being that kind of a dad for True, of being his baseball coach, and it broke my heart to know it was never going to happen, because I was never going to make it happen.

For two people who were obsessive planners, Diana and I had never actually discussed the fact that we were going to have only one kid. It just went unspoken. The most she'd say was that she'd never do it again unless I was around more often. I'd insist that "you and True are the most important things in my life." I'd say it because I *wanted* to be that person—but I wasn't. Family didn't silence the invisible storm in my mind the way work did. And lately even work wasn't cutting it. I'd built up too great a tolerance for that medicine. I needed something more—much more.

DIANA

I knew what would happen if he ran. I knew True and I would never see him, I knew I'd have to put my business on pause and figure out how we would survive financially for the next two years. But the whole thing felt much bigger than me. I believed that Jason had the best chance to beat Donald Trump—and compared to that, what did my own comfort matter? I'd worked so hard to build a business as a speaker and consultant but I was ready to kiss it all goodbye, because I believed as much as he believed.

The thing that gave me the most peace was reading the biography of Coretta Scott King. I hadn't realized that she had been a great singer and performer before Martin Luther King Jr. came along. She'd given up her career because she saw what her husband was capable of. She didn't just endure the death threats, the arrests, the beatings and the bombings; she maintained poise and composure and was the north star for her family, helping them see that they had a responsibility to the greater good.

It's not that I compared Jason to Martin Luther King Jr., but it was comforting to read the words of someone who understood what it was like to give up personal dreams for a bigger cause and maintain some semblance of normalcy under constant media scrutiny. I didn't know anybody who had been through this. I kept her book under the bed. I read it constantly.

But as comforting as it was to know that another person had been in an even more difficult situation, it didn't help my anxiety—it made it much worse. I was mentally preparing for Donald Trump's negative propaganda machine and what it would do to Jason and me and our family. I was terrified.

I'd met Barack and Michelle Obama once before, at a party in the summer of 2017. A mutual friend introduced us, but the president cut them off. "I know who this is! Jason, you and I need to be working together!" I talked with the two of them for about fifteen minutes, and the president and I departed that event with the intent of sitting down to visit soon. Not in a million years did I think he really meant it.

And now, here I was in his office, bonding with him over life, family, politics, Afghanistan, writing (we were both working on books), and mostly—to my surprise—my future. He listened actively, and he was generous, but not presumptuous, with his advice.

We laughed so often, I felt like I was hanging out with an old friend. They say you should never meet your heroes, but mine didn't just live up to my expectations—he blew them away.

I kept my cool for over an hour, but when Obama summarized our conversation, my expression likely betrayed me. He ran through the strengths of other potential presidential candidates in the race: experience, name recognition, donor networks, and obvious qualifications for the job, gently reminding me I lacked all of these advantages. He leaned forward, elbows on his knees, and looked thoughtfully down at the floor. Then the most admired man on

Planet Earth looked me straight in the eye and said, "But Jason, you have what I had. You're the natural."

Minutes later, he walked me to the elevator, but I'm not sure that my feet touched the ground until it was time to go home.

Somewhere over Kentucky, I realized what a gift he'd given me. To hear Josh and Donna from *The West Wing* tell me I could be president was one thing. To hear that the greatest president of my lifetime thought I could do the job was something I'd only dreamed of.

Shortly thereafter, high-profile alums of Obamaland began reaching out to Abe as well. We (and, I'm sure, a few other potential campaigns) were getting the full benefit of his mentorship.

A few weeks later, the small list of prospective candidates brought in for a meeting with President Obama was ferreted out by *Politico* and picked up by every major outlet. My name, obviously, was on the list.

From that point forward, instead of struggling to be taken seriously by the richest donors in the party, skepticism about my candidacy among the "donor class" was practically erased. In fact, some of the heaviest hitters in the party had switched from dodging my calls to requesting meetings. I told my family I was running. Abe and I got ready to mobilize the armies we'd built in Iowa and New Hampshire and Nevada. Diana steeled herself. I wasn't in this thing to make some speeches and get my message out there and be a bigger player in the party. I felt I was the best hope we had to beat Donald Trump. I was in it to win.

8

I DON'T WANT TO
DO THIS ANYMORE

If I'd been moving fast before, now it was like I'd been fired out of a cannon, and there were only two options: keep flying or crash. Literally, I was in the air all the time.

Every Monday at 2:30, I had to join a scheduling call. The number of people on the call had grown to fifteen, each representing something in my life that needed my attention: Let America Vote, *Majority 54*, fundraising for the presidential run, and on and on. Diana would be on these calls too: she represented Home. That was what my Home life had become: a thing to be scheduled. And it was the first thing I'd sacrifice.

We tried not to plan anything more than four or five weeks out, so at the top of every call, the calendar would have a block of days about a month away where nothing was slotted in. I'd always look at that empty block and say, *That's when I'll be home. That's when I'll be with my family.* And then one by one I'd watch as people looked at that beautiful empty space and said, "Okay, let's schedule it then." By the end, it would be booked solid. I never complained, but I felt awful every single time: I'd let down my family. Again. It was the lowest point of every week. Staffers dreaded having to ask me for anything on Monday afternoon after that call.

My paranoia about my family's safety reached new levels too, and my limited time at home was often spent arguing with Diana about how far ahead she let True get on his bike during walks or what route they took to school.

I'd also given up on sleeping. In the past, I constantly searched for new sleep hacks, trying anything; now I stopped trying. I told myself I just didn't need much sleep. The night terrors had always happened every night, but now they happened *all* night. During the day, there was this constant constriction in my chest, like my shirt was too tight.

I knew that what I was doing was bad for me, but I decided I had to do it because it would be good for others. I had to run, and I had to run *like this*, because to me, only two questions mattered: Do you think you're the best person to be the commander in chief? And do you think you're the best person to beat the worst person ever to become commander in chief? If both of those answers are yes, what choice do you have?

For more than a year, I'd been able to distract myself by taking on bright, shiny new challenges. They gave me endorphin boosts. But now those things weren't as challenging anymore. I was tired all the time. I *hurt* all the time. It was time to up the dose again.

In April 2018, "My Generation" blasted from the speakers as I jogged onto the stage at the Radisson Hotel in Nashua, New Hampshire. With a giant American flag looming behind me, I looked out over the packed room, where six hundred Democratic power brokers were in attendance. Each had been provided with a small bottle of Kansas City's Gates BBQ sauce and a set of Let America Vote thundersticks. This was the McIntyre-Shaheen 100 Club Dinner, a mouthful of a name for the most important night in New Hampshire Democratic politics. In the two prior years, the keynote speakers had been Hillary Clinton and Joe Biden; in 2019 it would be Elizabeth Warren. On this night it was Jason Kander.

Diana and True were there in the ballroom—True's lasting memory of the event was the fact that he got thundersticks. Also in attendance were Stephen, the mayor of Kansas City, the future mayor of

St. Louis, and several other prominent Missouri Democrats. They'd flown to New Hampshire because they understood this was a momentous evening, and they wanted to be there to support me. Nine staffers from Let America Vote were working the event, and our traveling videographer was on hand to capture every second.

My parents and grandfather were watching it live on C-SPAN's *Road to the White House 2020*. Everyone at that dinner knew the machine I'd constructed in their state, and every single person knew why I was there: to convince them I was presidential material. And they, in turn, were there to decide whether I was the real deal.

Over the previous year, I'd been following a schedule that made my Senate campaign look like a part-time hobby. I boarded more than three hundred airplanes, headlined major Democratic Party events in forty-six states, made a dozen trips each to Iowa and New Hampshire, built a vast network of donors, put together a senior campaign staff in waiting, and—with the help of Brad Whitford—reunited the entire cast of *The West Wing* to cut an ad for Let America Vote. All this, yet I had formally announced only that I "would consider running after the 2018 midterms." That was bullshit. I was running, and Abe and I made sure everyone who mattered knew it.

Running for president in the early states has little to do with big speeches. It's tiny living rooms with a precinct captain's dog in your lap and snowbound coffee shops with a few moms holding a book club in the corner. In terms of day-to-day campaigning, the best training for the initial part of running for president hadn't been my statewide races; it had been knocking on doors in Kansas City in my very first campaign.

I excelled at these intimate gatherings because I liked doing them. I'd always drawn my energy from other people, and to walk into a room where I knew my job was to convince a jaded Concord schoolteacher that I should be the president was the kind of challenge—at least early on—that could get me excited.

This, however—this was different. This was the speech that every state party dinner and every town hall had been leading up to. The political parties in New Hampshire—Republican and

Democrat—were acutely aware of their power. Their state motto is "Live Free or Die," but it might as well be "Vote First or Die." Whatever I thought of a state that was over 90 percent white maintaining this kung fu grip on the anointment of candidates for the presidency, I had to respect the seriousness with which they took their role. The Democrats were far from sold on the thirty-six-year-old who'd never been elected to an office higher than Missouri secretary of state. If this speech was anything less than a barn burner, I'd be toast.

Twenty-seven minutes later they were on their feet, stamping and shouting and beating those thundersticks so hard, I thought the roof was going to come loose. The selfie line afterward took forty-five minutes. One veteran reporter later told Abe that this "doesn't happen in New Hampshire this early." It was the speech of my life, the high point of my career so far, and I'd fucking nailed it. I wasn't just feeling good. I was euphoric.

The next morning at the airport in Manchester, the TSA guy examining my ID called me "our next president." But minutes later, as I boarded the plane with Diana and True, something had changed. I realized I felt *nothing*.

The endorphins were gone. The numbness, the emptiness, and the self-loathing rushed back in to take their place. I felt as dead as I had on November 9, 2016.

That was the first time I understood that something was seriously wrong with me.

Sometimes the journey to rock bottom begins at the very top. For me, the trip down even included a layover in paradise.

After New Hampshire, Senator Brian Schatz arranged for me and my family to come to Hawaii to give a speech, and the Hawaii Dems put us up in a resort for a few days. To my surprise, I felt safe there, safer than I had in a long time. I slowed down. I ignored my phone. I ran on the beach. I taught True to swim. I lay next to Diana while we both read books.

I tried to take stock. I realized how exhausted I was. I'd been living like this for so long that I didn't remember it hadn't always been this way.

The sensation of being constantly in danger wasn't gone, but it was softer. It had been so, so long since I had felt anything but guilt and fear, so long since I'd let my guard down.

Abe was on the trip too, and one evening when Abe, Diana, and I were at dinner, I quietly confided that I didn't think I had it in me to keep going. Over the recent months, I'd been hinting at how exhausted I was, but lately things had gotten worse. I was depressed, and I had even said as much directly to Abe a couple of times.

I was numb to the highs of the campaign trail. It was like I was having an out-of-body experience, watching Jason Kander walk into donor meetings and give speeches. I was having more and more trouble concentrating. Maybe it was the sleep deprivation, maybe it was something more, but I was beginning to have thoughts that frightened me. It wasn't that I was suicidal, it was more like I was coming to understand why some people chose suicide. All I knew was I didn't want to feel like this anymore. I was afraid of what might happen to me if I kept going, but I was even more terrified of what might happen if I stopped.

"Okay, so what if I didn't run?" I sheepishly asked. And held my breath.

Abe thought for a second. "You could just go home and run for mayor of Kansas City."

I felt an immediate and powerful wave of relief. Abe and I had discussed, in passing, such a prospect before the presidential run became a real possibility. I had thought of being mayor as a really cool thing I could've done, had I not ascended so quickly.

But now the idea was a revelation. I didn't need much convincing. Mayor was definitely what I needed to do. It seemed like the perfect solution.

Diana, Abe, and I spent the final two days of the trip working on logistics. Diana started playing around with ideas for the campaign logo.

Diana

Abe worked so hard to try to take care of Jason. He and I would frequently text behind Jason's back about what we thought he needed or how to protect him from information that we thought might bother him.

Abe had moved his whole life to Iowa, and now, on a moment's notice, he packed up his stuff and, along with Kellyn, moved to Kansas City for the mayor's race. He even mentioned that night at dinner how interested he had always been in city issues. That might have even been true, but it was clear to me that he was just trying to support Jason the best way he knew how. I think he and I were both scared of what would happen to Jason if he just stopped.

In all the time we spent planning over the next two days, no one mentioned Jason's mental health. After the issue came up during that one dinner, it never resurfaced. We all quickly moved right back into what felt comfortable to us—planning a campaign.

———

I came home to Kansas City with a two-pronged plan to fix whatever was wrong with me. Part One was to win the mayoral race. I was trying a whole new tactic in my fight for redemption: serving my neighbors in my hometown, striving to make really tangible change that I could see. I thought perhaps that was the missing element in my life; perhaps that would be the validation I needed in order to feel I'd done enough.

Part Two was to start getting help at the VA—I didn't think I had post-traumatic stress disorder, but clearly I needed some kind of therapy. Dutifully, I started to fill out the VA paperwork but then got freaked out by what it might mean if my honest answers to the screening questions became public. So I minimized my symptoms and sent in the documents. Weeks later, I got a letter in the mail. My request for help had been rejected. Running for mayor would have to do.

When Abe and I told the campaign staff about this major change, they were understandably shocked and confused. Some were furious—they'd joined up to run a presidential campaign, not a mayoral race in a midsize city. I didn't dare tell them the real reason for this new direction. Lamely, I said I didn't think we could raise the money for a presidential run. That especially—and justifiably—pissed off Kellyn, who was now my finance director. She voiced her disagreement: "Uh, yeah, we definitely can." I just shook my head and apologized again. I was doing the very thing I had spent years avoiding: letting down people who believed in me. The guilt was searing. I wanted to vanish, to dissolve. Still, to a person, my staff remained loyal and ready to lower their sights alongside mine.

On July 14, 2018, in an office above the Negro Leagues Baseball Museum in Kansas City, I was going over my notes for the announcement speech. The hall downstairs was packed, and the crowd was among the most diverse I'd ever seen in my town.

I looked at my opening line: "So, almost 140 years ago, this guy named Felix Kander came to Kansas City, and I think he would be pretty proud to know that his great-great-grandson can stand here and say, 'I'm Jason Kander, and I'm running for mayor.'"

It was a good line. I planned to deliver it while rhythmically whacking the lectern for emphasis. I'd sell it completely, and the cheering would start even before I finished. But right now I was alone, trying to convince myself that this entire campaign wasn't the result of pure cowardice.

I was telling myself that this was what a good person would do. That this was the right thing to do. This had always been my spiel as a politician—from state to state, I'd stood on stage and insisted that politicians always knew the right thing to do; the only hard part was *doing* it. Now here I was, soaked in doubt. I wanted to believe that this *was* the right thing—why did it feel so empty?

I hadn't stopped believing I was the best hope to defeat Donald Trump in 2020. But here I was, backing down from that duty. I was retreating. I'd let myself be beat by whatever was wrong with me. Shame and guilt rose like bile in my throat.

But I did recognize something—just a glimmer of relief telling me that perhaps I'd made a wise choice. What if I'd gotten fully into the race, won the nomination—and then collapsed midway through the campaign, leaving the party rudderless? Here, I told myself, I had a chance to serve my city, and service would make me feel better. Maybe I didn't feel good now, but I would—once I'd gotten people housing, once I'd cut down violent crime.

The front page of CNN.com said it all: "Rumored 2020 candidate Jason Kander is running . . . for mayor?"

This was the first campaign in which I knew from day one that I was going to win. The day I announced, we sold $25,000 worth of campaign T-shirts. Within the first few weeks, my first book—which I'd written almost entirely at thirty-five thousand feet while crisscrossing the country—hit the *New York Times* bestseller list. My podcast, *Majority 54*, remained among the most downloaded in America, and I talked about my plans for Kansas City everywhere, from *Morning Joe* to *Late Night with Seth Meyers*.

Instead of struggling for name recognition like everyone else in the race, I had nearly 100 percent *face* recognition. People would slow down as they drove past me on the street and yell, "I'm voting for you!" When I knocked on people's doors, they'd come out holding my book and ask me to sign it.

Now that I was out of the presidential race, scores of presidential candidates were clamoring to host fundraisers for my mayoral campaign. Winning my support had become a coveted prize. A few even offered to fly to Kansas City and join me as I knocked on doors.

It was downright unfair and—by anyone's objective standard—the ideal way to run for office.

And yet I spent that whole campaign in a constant state of rage.

I don't mean the kind of righteous fury that fueled me through my race for the statehouse years earlier. I mean, I had a full-blown anger problem.

I was terrible to be around. I went to bed angry and I woke up angry. I was impervious to good news and virtually impossible to cheer up. I yelled at Diana. I snapped at staff, something I had never done before. I wanted to scream at voters: *Oh, you want to talk about tax credits? Who gives a shit? People are being shot in the fucking street.* All I cared about was Kansas City's murder rate. I felt responsible for it—for every single murder. It didn't matter that I wasn't mayor yet; I was going to be, and already I was failing to protect people. I read obsessively about combating urban violence and talked to mayors around the country about the strategies they were using.

I was a soldier running for mayor as if it were sheriff.

Of the ninety-nine days I spent on that campaign, that rage cooled on one day only: when Kellyn, who was managing the mayoral campaign, and I toured a facility called the Veterans Community Project (VCP). VCP provided housing for homeless veterans as well as a walk-in clinic for any who had slipped through the cracks of the system. They'd built a village of tiny houses, designed to replicate the layout of base housing—that is, the last place where these veterans had experienced stability and success. Founded and run by veterans of Iraq and Afghanistan, VCP had a remarkable vibe. It was as if a forward operating base in Afghanistan and a startup in Silicon Valley had had a baby.

The place inspired me. When I got home that night, I grimly joked to Diana, "I wish I could just quit everything I'm doing and go work there."

In the fall, the polls showed that I had a commanding lead. We shattered fundraising records, we had a huge waiting list to host house parties, and I had the endorsement of nearly every prominent person in Kansas City. And the whole time I was thinking more and more about ending my life. I had reached a point where I hated myself so fiercely that I saw myself as a burden, and I was convinced Diana and True would be better off without me.

DIANA

Jason would frequently say to me, "I feel like I'm disappointing everyone." I used to think he used that line to get out of whatever I

had just asked him to do. But more and more he would say it with-
out being prompted. Then one day, as we were in the third hour of
writing thank-you cards to people who had preordered Jason's book,
he said: "Do you ever think the world would be a better place without
you in it?"

"No," I said. And then I asked, "Are you talking about suicide?"

He sheepishly nodded as his eyes filled with tears. I felt a wave of
emotion: sadness, anger, desperation.

When you read those memes online—"If you're thinking about
suicide, know that you are loved and can always ask for help"—it's
easy to think, Oh, that's definitely how I would respond. *But in-*
stead I handled it in the worst way possible. My imagination flashed
to Jason's funeral—having to comfort his parents, explaining to True
why his dad wasn't coming back, raising True on my own. And I
lost it. I cried, then got mad, and the whole thing ended with Jason
having to comfort me.

I had no training in how to talk to someone who tells you some-
thing like that. I had no idea how to cope—all I knew was that I felt
scared and powerless, and I couldn't tell a soul.

People often ask me what finally caused me to get help. Well, I'm
about to tell you how it went down, but you have to understand that
if it hadn't happened this way, it would have happened some other
way, a few days or a few weeks later. Just like there was no single
moment of trauma that caused this trouble in the first place, there
was no single event that caused me to reach for real help.

For a few weeks, an obscure notion had been circling in my mind:
I was at some sort of crossroads. It often manifested as a line from
the movie *The Shawshank Redemption:* "Get busy living or get busy
dying."

One night at the end of September, after yet another stressful
day of simultaneously running for mayor and leading Let America
Vote, I felt I had hit a new low—a sense that whereas things had

been getting steadily worse for years, now for a few months they'd been getting worse even faster, which was frightening. Sitting next to Diana on the couch in our living room, I was struck by the idea that it was time to try *anything*.

I was still holding on to the idea that perhaps I could stop the problem where it was—or at least escape the increasingly common feeling that if I stopped existing, things would be better for everyone. This thought process—like a tiny seedling of hope sprouting through a crack in the pavement of depression—had been percolating for a couple of weeks. That's why I'd already looked up the number for the Veterans Crisis Line.*

I didn't want to die, or at least I didn't want to want to die, and I vaguely understood that I should take these feelings seriously. I decided I'd try again at the VA. This time, instead of filling out an online form, I'd just call, even though I assumed they'd tell me to go away. I was really timid about it. I went to our bedroom to make the phone call in private. *I want to talk to someone, but I'm not really in crisis*, I told myself. *I'm just confused and looking for someone who might understand, someone who might have an idea of what the heck I need to do.* I felt guilty for taking up time on the hotline that could've been used for someone more worthy. As the phone rang, I felt like an imposter.

The woman on the other end of the phone line took me by surprise with one of her first questions: "Have you had suicidal thoughts?" she asked.

I had never acknowledged this to anyone except Diana. I said yes. I expected the woman to be shocked. She wasn't fazed at all. She just asked me to walk her through it, to tell her where I served, how I was feeling. I started crying—just saying the words out loud was like shattering the glass and pulling the fire alarm and setting off the sprinkler system. She replied calmly, "Okay, you're going to need to head into your local VA and get enrolled in the system. We'll call and check up on your progress in the coming days."

I was floored by the tone of her voice. Apparently I was just like

---

* 1-800-273-8255. Take a moment and save it in your phone.

anybody else she talked to that night, just like all those vets I'd spoken to over the years.

I hung up the phone, walked back out to the living room, and immediately googled "post-traumatic stress disorder." This time I read the information with an open mind, not just to prove to myself I didn't have it—and it was like the description had been written about me.

Standing in the kitchen, leaning on Diana, I cried. Hard. Like my entire body was a wet rag and someone was just twisting and wringing every drop of grief out of me.

For a long time that evening, Diana just held me. I lay there on the couch, my head in her lap, staring at the ceiling, wondering what the hell I was going to do. What *we* were going to do.

I had been hurt over there. I had been wounded. And all this time, I didn't know.

And that's when I finally said, "I don't want to do this anymore."

9

JASON KANDER? HE DIED.

You have to be a little crazy to be in politics, but what you *cannot* be is mentally ill. Or at least, that's how it seems if you look back at the past few hundred years of American government (which started as a rebellion against a king who *was* mentally ill). Even a suspicion that a candidate or someone in government has a psychological problem is a death sentence in politics. Just look at Thomas Eagleton. He was a political phenom—Missouri attorney general at thirty-one, US senator from Missouri at thirty-nine. You might call him another young vet from the Midwest. In the 1972 election, George McGovern tapped him to be his running mate against Nixon.

Then, two weeks after the Democratic Convention, the news broke: Eagleton had been hospitalized multiple times for electroshock treatments for clinical depression.

At first, McGovern claimed he would stick with Eagleton "1,000 percent." Six days later, he asked Eagleton to withdraw from the ticket. This wasn't cruelty—prominent psychiatrists, including Eagleton's own doctors, had told him that the depression could return and thereby put the country in danger. The episode allowed Republicans to claim that McGovern—already painted as a wild-eyed leftie—had crappy judgment. He lost everywhere except Massachusetts and DC. One Democratic strategist called Eagleton "one of the great train wrecks of all time."

Attitudes toward mental health have changed since then (Mike

Dukakis, who also got crushed in a presidential election, is now an ambassador for the value of electroconvulsive therapy)—but they haven't changed *that* much. As much as I knew I needed help, I was after all a politician.

The election was just a few months away. I didn't know what treatment entailed, but I knew I couldn't do it and run for mayor at the same time. My daily schedule was filled from sunrise to 10 p.m. with meetings, speaking events, and call time, and ever since I'd hung up the phone after talking with the woman from the Veteran's Crisis Line, I'd lost the energy and the desire to do any of it.

I finally had to admit that the story I'd been telling myself for a decade, that *I'll feel better when* . . . was a lie. Winning an office had never made any of it any better, and being mayor would be no different. I had no idea if I was even capable of getting better—if the damage was permanent—but at long last, I was ready to throw my entire self into finding out. I'd finally arrived at Rock Bottom, the international capital of having zero fucks left to give.

Two days later, I walked into the VA for the first time. I answered all the questions again. I met the psychiatrist who assumed I was hearing voices. The great creaking, lumbering machine of the VA began to turn its wheels. That was—I thought at the time—the easy part. The next part was harder.

Once again, it was time to write a letter. I decided not only to announce that I was quitting the race and stepping away from Let America Vote but also to publicly share the reason.

In my announcement, I unloaded everything I'd been hiding from myself and from the world for eleven years. I admitted that I had post-traumatic stress symptoms and suicidal thoughts, that I was ending my campaign and stepping out of public life to focus on getting better. I vowed to come back stronger, but part of that was me trying on the Jason Kander suit one last time. I *hoped* I'd come back. I didn't really believe it, though. I was throwing away everything my team and I had built—the only part of my life that was going well.

Before he hit send, Abe looked at me and asked, "Are you sure this is what you want people to remember about you?"

"I need there to be some good that comes out of this," I said.

Abe and I hugged.

"It sucks to feel so weak," I said, tears in my eyes.

"There's nothing weak about what you're doing," said Abe.

I thanked him for his friendship, then went to the bathroom to take a long, sad shower. Abe sat down on the living room couch and hit the self-destruct button on my (and to some extent, his) political future.

At the exact moment we posted the letter, Diana happened to be in Ohio, giving a speech she hadn't been able to get out of. She raced through the presentation, sick at heart, trying to get off the stage before the news broke and people started asking her about it.

True, meanwhile, was at his pre-K program. All my thoughts went to him. For the rest of his life, people would say to him, "Oh, I heard about your dad, I hope he's doing okay," and just in case he remembered this day, I wanted that memory to be a happy one.

Stephen drove in from Columbia, and we spent much of the afternoon assembling a giant Pop-a-Shot basketball game so it would be ready when True got home. Meanwhile I entrusted my phone to Abe, who sat on the floor, his back to the wall, fielding a deluge of calls and texts from friends and reporters.

Abe was an absolute brick wall that day. Nothing got through him. Nancy Pelosi called, and I had no idea. Some friends, like Jake Tapper, were so persistent, they wouldn't give up until they'd heard from me to make sure I was all right. Joe Biden texted several times to offer support and encouragement. Elizabeth Warren spoke to Abe for quite a while, then volunteered—without being asked—to raise funds for Let America Vote while I was out of commission (which is exactly what she did). I was vaguely aware of this outpouring of support, but I was also determined to ignore it and, for once, be as present as possible. That's why I gave Abe my phone and asked everyone around me to say nothing to me about the public reaction.

Most reporters who called, asking to talk to me, gave the decent response when Abe told them no, but there were a few exceptions. One local reporter demanded an interview. Abe said no. He'd said no to *Good Morning America* a minute earlier, but this local guy was insistent.

"Don't you think he owes it to the people to talk about this?" he finally barked.

"You know what?" said Abe. "Go fuck yourself." And he hung up. That one I overhead. I think I even smiled. Abe had been wanting to do that for years.

To the outside world, my silence probably gave the impression that I was in such a bad way, I was incapable of sitting for an interview. But really, what could I have said? I didn't yet know anything about PTSD, so who was I to speak publicly about it? I'd come to accept that I had it only about seventy-two hours earlier. That would be like getting a cancer diagnosis one day, and the next day appearing on *60 Minutes* to explain your illness to the entire country. If anything, I was maybe the *worst*-qualified person to talk about PTSD—I'd spent eleven fucking years being wrong about it.

But there was another reason to avoid the media at this point. I'd understood for years that public adulation and media attention had been a useful distraction for me. As long as I was on TV, on radio, doing interviews, listening to other people's voices in my ear and firing back, then reading the fawning comments on social media, I wasn't alone with myself.

So, I had figured out exactly one thing: I needed to focus on healing, and it would require my full attention. I had no idea how big or small a story my announcement was, and I was trying hard not to care.

Stephen and I were having trouble putting together the Pop-a-Shot. Between phone calls, Abe tried to help, but we were not a handy bunch. The instructions said that assembly should take no more than an hour, but it took us the whole damn day.

It was almost time to pick up True at school when I had a jarring thought: would I be allowed to pick him up by myself? I'd admitted to the world that I had suicidal thoughts. What if I was prevented from leaving with him?

The full impact of what I'd done set in then. It wasn't the calls or the messages. It was the idea that suddenly I'd gone from being the next mayor of Kansas City to being a guy who'd sunk so low, his kid's school might not feel safe handing the boy over to him.

Abe and Stephen walked with me to the school, and in the end, it wasn't a problem. That evening, my dad and my brothers Jeff and Mel came over, and brought along True's cousin, whom he adores. The Pop-a-Shot didn't fall apart, Diana came home, and—if nothing else—I had given True a pretty fun afternoon. That small accomplishment, combined with having people I cared about by my side all day, made for a glimmer of hope.

As I went to sleep that night, I swam through strange seas: grief over the loss of everything I'd built and relief that I no longer had to build it. Then one fear bore down on me like a giant wave: *What if it's too late? What if I can't get better?*

DIANA

Over the previous ten years, Jason and I had gradually become two people in a marriage who both felt alone in their pain. Starting with his deployment to Afghanistan, we shared less and less of how we were feeling in order to protect each other, until one day we just didn't share anything at all. We were quietly living isolated lives and feeling desperate.

PTSD tricks you into feeling that there is something wrong with you, that no one will understand what you are going through, that they will judge you for your dumb thoughts, that they won't like you anymore. That includes family and significant others. It makes you think that you're protecting yourself, but really, you're left all alone with your intrusive thoughts swirling loudly inside your head.

Within days of Jason's announcement, we received thousands of letters and emails from people who had never told anyone about their own dark thoughts and isolating behavior. Friends whom we had known incredibly well confided their dark times, their thoughts of suicide, and even their attempts. A volunteer on Jason's mayoral

*campaign shared that he had been contemplating suicide the morning
of the announcement; reading Jason's words made him pause and
reach out to his mom. Strangers shared their struggles, and so did a
lot of people we knew personally.*

*I realized that it wasn't just me, and it wasn't just him. We felt a
level of connection like never before. Not just to each other, but to our
community. We heard from so many people that I felt I could walk
into any room, any restaurant, and safely bet that at least a third of
the people in that space were dealing with a serious mental health
issue in their family. That was incredibly comforting, but it didn't
change the uphill battle Jason was facing.*

*Making his announcement didn't alleviate his condition. In fact,
it took away the only safety net he had. Without the activities that at
least somewhat suppressed it, Jason's depression intensified. He kept
asking me if I thought he could ever be happy, and I was hoping the
answer was yes. But I honestly had no idea.*

———

I'd like to say that the day after my announcement was when the
real work began, but in fact it was the start of something else: stum-
bling around in the VA's labyrinthine bureaucracy. After my visit the
day before, I was told that I wouldn't be able to see a therapist for
months. Moreover, my service had taken place eleven years prior,
and the VA had a stupid rule that if your combat deployment had
ended five or more years earlier, you had to effectively prove that
you had trauma. So now I was fighting an administrative case to
prove I had the condition I'd denied to myself and the outside world
for years, and I had no idea where to begin.* I knew that for vets,
access to benefits was messed up, but I didn't know *how* messed up:
here I was, with a phone full of influential contacts, a Georgetown

---

* In the past couple of years, the VA both repealed the five-year rule and changed the
benefits process so that applicants can receive mental health treatment while their
application is considered. This wasn't the case in 2018, however.

law degree, and high-level government experience, and I was overwhelmed by the VA system. How in hell was anyone supposed to navigate it?

The answer—to this, and to so much else—was a guy named Bryan Meyer. I'd met him back in 2012, when he volunteered on my campaign, and we'd stayed in touch. Bryan was running the Veterans Community Project, whose campus I'd toured during my mayoral campaign. Before my announcement came out, I'd texted Bryan to let him know. He got right on the phone, and I explained how confused I was by the whole VA process. He suggested I come down to his office. So, six weeks after my VIP tour of VCP as the future mayor of Kansas City, I walked through the doors of the outreach center—just like thousands of KC veterans before me.

I'd loved VCP the minute I first saw it. Everything about the place was pure genius. For years, people had been trying to understand why homeless veterans strongly resisted staying in shelters, but VCP's founders knew it was because untreated PTSD makes it impossible to sleep around strangers.* They designed tiny homes specifically for people with PTSD (the bed faces the door, the windows never look into another unit, and so on). The village was laid out to replicate base housing, which allowed residents to restart the transition from military to civilian life. Unlike the process of engaging with most providers of veteran services, no red tape was involved, and every type of veteran, with every type of service or discharge status, qualified for 100 percent of VCP's programming, including its one-of-a-kind case management system. Bryan and the other veterans of Iraq and Afghanistan who had maxed out their credit cards and taken out second mortgages on their homes to create VCP had achieved a reduction in street homelessness among veterans in Kansas City that was unmatched anywhere else in America.

I wasn't the guy running for office anymore. Now I was just a vet who needed support, and the folks at VCP instantly began treating me like one of their own. They helped expedite my paperwork.

---

* This, it turns out, is why I'd never been able to sleep on red-eye flights.

Bryan, like all the guys who created VCP, was both a combat veteran and a veteran of the Kansas City VA's PTSD clinic, so he and his fellow cofounders suggested some therapists at the VA whom I might work well with. By the time they were done, my first appointment was scheduled.

That week, and the weeks to come, even after my therapy sessions began, were a gray, sullen haze. For almost half a decade, my life had been scheduled down to the minute, which allowed me to dodge myself and my mind for long stretches of time. Now, all at once, that schedule had been wiped clean. I went from stump speeches and fundraisers to attempting to be helpful around the house; of course, I had barely any idea how to do that. I couldn't cook, and all I knew about laundry was my shirt-in-the-dryer trick. For years I'd neglected housework for the same reason I'd neglected my body and my mind: it wasn't as important as the big thing I was doing. For six years I'd had an entire pit crew tending to me as if I was a race car, but now I found myself standing on the side of the road with a smoking engine, a half-empty can of motor oil, and a rusty wrench. And I was just wrecked by depression.

Opting for a lower public profile had done nothing to lessen the feeling that danger was ever-present. On a rare family outing, we stopped to get some gas. As I was filling the tank, Diana and True went inside to use the restroom. I tried to peer through the windows to make sure they were okay when a slight man in his early twenties, dressed in an oversized flannel shirt and jeans, approached me from behind. He was carrying a gas can. I gave him a look that made it clear he should stop walking toward me. He meekly explained that he and his girlfriend had run out of gas, and they had no money. He asked if I'd mind filling up his gas can. I looked past him and saw a woman about his age sitting on the curb next to an old beat-up minivan.

I am a generous person, which is the kind of thing someone says right before telling you a story about how they've been anything but generous, but it's true. Over the years, while traveling for work, I've given away the literal coat off my back to a homeless person, handed out food, bought meals . . . but this was different. My family was inside a building and I was outside; to me it looked like an obvious trap. Adrenaline took over.

"I'm sorry, I can't help you," I said. "Please step away." The man went to talk to someone at another pump, and though my tank was only half full, I jammed the gas nozzle back into its holster and moved with purpose into the store. Diana was still waiting in line to use the restroom.

"Time to go," I said.

"But we've been waiting this whole time—"

"You'll go to the bathroom at the next stop. We're leaving now."

Diana could tell from my tone that this wasn't the time to ask questions. I loaded everyone into the car and sped off, then re-counted what had happened. She kept asking questions, trying to figure out what had bothered me so much. It was hard to explain. I was sweating, my heart was racing, and I was gripping the wheel like it might try to run away. In my mind, this had been an incident, a close call. Someone had tried to harm us, and thanks to my quick thinking, we'd escaped.

Forty-five minutes later, Diana had persuaded me to consider the possibility that the man with the gas can might have just been a man who had run out of gas and money. Intellectually I realized that it was probably true, yet my whole body and most of my brain rejected the idea completely.

The weeks went by slowly. Even in my bubble, I began to under-stand that my withdrawal from the mayoral race had made the news. Some days I'd be feeling tolerable, and then I'd go to a grocery store and someone would recognize me and try to console me, but I'd wind up consoling them. As it turned out, admitting I'd felt suicidal meant everyone I knew—and many people I didn't know—suddenly felt it was their responsibility to make sure I didn't kill myself. I'd be minding my business, picking out fruit in the produce aisle, and a stranger would lean in and whisper "The world is a better place with you in it." This was kind but also mildly creepy and majorly awkward.

One night, Diana and I went to a David Cook concert. I knew David Cook—a Kansas City talent—and he'd invited me months ago, but I had declined. I knew I'd be too busy with the mayoral race. Now I reached out, asking if there was any way he could get me in.

It was the first night out that Diana and I had had since the announcement, and we were invited backstage after the concert. I knew it was going to feel weird to rub elbows with shiny people again, but I hadn't quite factored in what that might entail. Literally *everybody* was cautiously asking, "Heeyyyyy . . . so how you doing?" They seemed shocked to see me at something as frivolous as a concert. One person, who claimed not to follow politics at all, told me she was a huge fan. "I work in mental health," she said.

Diana observed that I might be the "most famous depressed person in America." That was a hell of a pivot in my personal brand.

I needed people around me who didn't treat me like a time bomb. Thank God for Stephen. Other than my therapist, no one in my life in that moment better understood what I was going through. Stephen was—in so many ways—my mentor through this process. He had gotten his own kind of help, and he was there to support me as I did the same. But most important, he could see the whole experience as fodder for comedy. Laughing was proof I wouldn't crack, and dark humor actually gave me a lift.

When people saw me in a restaurant, they looked like they'd seen a ghost: "Oh my God, there's Jason Kander. He *died.*"

One day at the DMV, a guy came over and shook my hand. "Big fan," he said. Then he went and sat down, and I heard what he said as he pointed me out to the person next to him. And he didn't say, "That guy was secretary of state" or "I loved his book" or "I listened to his podcast." What he said was "That's Jason Kander. He just had to stop everything because he was *really* depressed."

I did my best to laugh, but it was hard not to let that stuff seep into me. I grew out my beard and started wearing a hat, which cut down on these exchanges because I was less recognizable—or maybe I looked less approachable.

• • •

Before seeing a therapist, the VA had me meet with a clinical social worker. The session was supposed to last half an hour—she would ask questions, I'd answer, and she'd fill out a form that would help determine whether I had PTSD, and if so, how severe it was. Our session lasted for two hours. Now that I had the chance to start talking about the problem, I couldn't stop. She almost smiled as she said, "Man, you are *ready* for therapy."

As I told her one of the stories that I couldn't get out of my mind—the one about Sabet and Hajji Zahir—my heart began racing, and I had trouble breathing. "You're reexperiencing," she told me. Instantly I knew she was right—I'd just never known the word for it. I'd written a sanitized version of that story in my first book, and one day, during my very abbreviated book tour, I'd actually told it on *Morning Joe*, and the same thing had happened. The trouble with breathing, my heart racing.

My clinical social worker had read my first book, which included a chapter that addressed the tics I experienced after my return from Afghanistan. I'd written: "I went and saw a guy about it and he gave me some helpful tips, like what not to eat before bed. I also declared a personal moratorium on movies and books about war . . . I researched post-traumatic stress and decided that's probably not what I was dealing with, but I also learned about 'battlemind' and the debriefings about it I'd never received from the Army." In that chapter, I settled comfortably on a diagnosis of "battlemind" and left it at that.

"Do you realize," the social worker asked calmly, "that you wrote an entire book to convince yourself you didn't have PTSD, and that nothing really happened over there? Why did you feel like you didn't earn it?"

"Because I have friends who were in combat, as in firefights," I said, "and I was just some asshole who went to meetings."

"Okay, but what do your friends who were in firefights say about what you did over there?"

"They always say they wouldn't have been willing to do it, but I just assume they're being nice."

"Okay," she said, "but those fights they were in? Yes, they were traumatic. But they lasted minutes, and those soldiers had a bunch of their brothers and sisters by their side. And when they came home, they weren't getting into more gun battles. Your meetings in Afghanistan, on the other hand, meant you and your translator went out more or less alone with no backup, no one even knowing where you were, totally vulnerable for hours at a time, in the most dangerous place on earth, to sit with people who might want to kill you."

"That's combat trauma," she continued, "no matter how you slice it. And when you came home, what did you do for a living? *You went to more high-stakes meetings.*"

Diana

Recognizing that you need help doesn't actually relieve the symptoms of your condition. Our decision to get a new dog, Talia, is a perfect example of that.

During the mayor's race Jason's PTSD symptoms were exacerbated, driven into the stratosphere. It affected me too. Online I saw people posting our home address, directions to our house, and where our son went to school, and it ramped up my fears about our safety. My response was to install two separate camera systems around the house; I put dead bolts on the bedroom doors; I added an alarm that monitored the doors and windows and any motion inside the house at night while we were sleeping. I did dozens of other things to secure the house, but none of it was enough. It didn't make me feel safer.

I started brainstorming with Jason about which former police officer or veteran we could convince to move into our spare bedroom, so that someone would always be there to protect True and me when Jason was traveling.

Then, after he dropped out of the race, this fear didn't cease. I felt we were in just as much danger. As I spent hours and hours research-

ing home safety tools online, I came across a dog-training facility in Kentucky that seemed to offer the perfect solution. I became a fan of its YouTube channel, admiring the magnificent skills these people were able to teach to powerful animals. They worked only with German shepherds, Doberman pinschers, and giant schnauzers. I was unfamiliar with the giant schnauzer breed, and I immediately fell in love. Bringing this type of dog into our lives felt like the perfect solution to a problem that had seemed unsolvable.

PTSD service dogs are truly remarkable. They can anticipate and lovingly disrupt panic attacks by detecting changes in your breathing or smelling your perspiration. They sleep by your bed, and when you have a nightmare, they jump on top of you and shake your body awake. And most important, they provide love and compassion. To deliver these therapeutic benefits, PTSD service dogs receive intensive specialized training. The nonprofits that provide these animals spend as much as $25,000 training each one.

Yep. We really should have gotten one.

But we didn't yet know anything about PTSD. What we *did* know was that the world was a dangerous place. Having a dog to protect our family would, we thought, make us feel safe.

In Europe, giant schnauzers are used as police dogs. They are a working breed, naturally protective of their people and suspicious of strangers. Talia was jet-black, huge, and strong as hell. Only five months old when she came into our lives, she had already been trained to deter, to guard, to assess potential aggressors, and—if necessary—to attack them.

The gentleman who sold her to us was familiar with my background, and he said to me, "She's trained just like you were—she's got those protection instincts."

That's right. Instead of a PTSD service dog, we got a dog with PTSD.

Soon after she came to live with us, it became obvious that Talia still needed some training, but not in protection or obedience. We needed to teach her etiquette. When I'd take her for a walk, she'd stay one step behind me on my left, just as I'd been trained to do when walking with a superior officer. Talia reminded me of a drill instructor. When someone would walk past us—on the other side of the street, no less—if they so much as made eye contact, she would bark menacingly, as if to say, "Are you looking at Jason right now? Don't you look at my Jason! Oh my God, you're *still* looking at him! You must be outside your mind right now!"

She walked with her head on a swivel, because she wasn't on a walk, she was conducting a presence patrol. And she didn't "walk" anyway, she swaggered, as if saying to herself, "Hell yeah, I *wish* a motherfucker *would.*"

But Talia was never aggressive toward the family. She minded us, including True. She clearly loved us, or at least she seemed to, because she was solely dedicated to protecting us. She would bark at anyone who dared walk in front of or behind our house. She slept—or rather, stood sentry—on the landing at the top of the stairs, where she could see all the doors. She was so focused on this duty, she didn't have any bandwidth left for affection. Anytime we'd try for a snuggle, she would have none of it. She didn't really care to be petted, and if you tried hugging her, she'd back up and politely snap at you to let you know she wasn't interested. It didn't feel like we'd bought a dog. It felt like we'd adopted Jason Bourne.

Diana decided that—in addition to continuing and enriching Talia's obedience training—she would help her to relax a bit. Her number-one goal became teaching this terminator, sent from the future by Skynet, to at least contemplate a cuddle on the couch.

Gradually, Talia settled into our home in Kansas City. So did I. The family had a lot of work to do, but at least we were finally together, and we had nothing but time.

10

THE MONSTER

Date of Note: Oct 09, 2018
Trauma History and Symptoms
DESCRIPTION OF VETERAN'S REPORTED TRAUMATIC
EXPERIENCES: "I didn't feel like 4 months justified getting help. It's embarrassing." Veteran discussed various
events but he noted almost being kidnapped as the most
re-occurring traumatic event. He also noted having re-
occurring images of a child he almost shot.

QUESTIONNAIRES: Vet completed the PTSD Symptom
Checklist (PCL5) and scored 72 out of 80 points. A score
above 38 indicates possible PTSD. The Veteran obtained
a score of 25 on the PHQ9 which suggests a severely depressed mood.*

My therapist's name was Nicholas Heinecke. He went by Nick. He
was tall and bearded, with a casual and welcoming air. His small office was adorned with patches and paintings clearly gifted to him by
other veterans. Most of the memorabilia was from the Vietnam era.
Nick was my age and knew who I was, but he'd moved to Missouri

---

* Any VA patient can obtain a copy of their full medical records. Notes made by my
therapist and his colleagues are quoted verbatim in this chapter.

only a year before and hadn't closely followed my career. He made a point of not googling me so that he could treat me like any other patient and deal with only the information I provided.

Right off the bat, Nick referred to PTSD as "The Monster." Together, he said, we were going to tame The Monster.

I told Nick my calendar was clear and taming The Monster was my focus, so I was available to come in every day. I figured the way to beat this thing was to work myself ragged. That had always been my MO.

"Yeah, I appreciate that," he said with a knowing smile. "But we're going to stick with weekly appointments because this will go better if we slow you down."

One of the first things Nick asked me was "What are your goals for treatment?"

"Well," I said, "I'd really love to be able to sleep through the night."

"Okay," he said. "What else?"

I stared up at the gritty drop ceiling. Most of that session, I kept staring up at the ceiling because my back had finally given out. I was lying on the floor with an ice pack—the same familiar position from all those hours of call time.

"I want the ability to be happy," I said. "I want to be able to feel good about myself without external validation."

It was almost funny. When I'd told the first VA psychologist that I thought I could be president, he'd thought I was crazy, but I'd fully believed it myself. Now I was saying I wanted to be happy, and that felt delusional. Ten years had passed since the last time I'd been able to feel good. Those happiness muscles had atrophied, perhaps beyond repair.

But Nick just nodded and wrote down what I said. "Good," he said. "Let's get to work."

Date of Note: Oct 18, 2018
Veteran completed session 1 of Prolonged Exposure therapy for PTSD.

Veteran discussed improved connection with family since dropping out of mayoral race. He voiced motivation for treatment and recognized the challenge treatment will present. He noted that the evaluation helped him feel validated about the symptoms he has struggled with over the years.

LETHALITY ASSESSMENT: When asked directly, Veteran denied any current or recent SI/HI.* Risk and protective factors were reviewed. Veteran remains future oriented. Risk of harm to self/others is judged as low presently.

When I started getting help, my great-uncle John said that "therapy is about getting a master's in yourself." His insight was spot on. Each of my sessions was split into two parts. First, I'd explain to Nick what I was feeling, describing my obsessive protectiveness and my inability to see myself as anything other than a failure. Then Nick would go to a whiteboard and sketch out diagrams showing how The Monster was causing those things. He was actively teaching me how my brain worked.

The second half was prolonged exposure therapy: Nick would have me turn on the voice recorder on my phone, close my eyes, and talk to him about a memory from Afghanistan as if it was happening in real time. This forced me to reexperience the moment while he asked me questions that someone who had never heard the story before might ask. Then, about forty-five minutes later, when I'd finished, I'd open my eyes and catch my breath. Then we'd rate how upset I was, using the SUDs scale of 1–100.†

Between therapy sessions, my assignment was to listen to the voice memo of the story. I wasn't allowed to multitask. I had to close my eyes and listen, with no distractions.

---

* Suicidal ideation / homicidal ideation.

† SUDs stands for "subjective units of distress." In our first session, Nick gave me permission to completely forget what SUDs stood for. And I did, until I looked it up for this book.

I had a lot of stories, but the one about Sabet and Hajji Zahir and the DIA guys leaving their weapons in the vehicle soon emerged as a good place to begin, at least to learn the technique.

The first few times I told that story, the same thing would happen: my heart would pound, my face would get flushed, and I'd clench my fists, unconsciously preparing for violence.

"Feel the feelings," Nick said.

I wasn't sure what that meant, but I nodded. Feel the feelings? It sounded like a cliché.

What I did appreciate about these prolonged exposure sessions— they served as a gateway to meditation. Some people find meditation difficult or even silly. I, however, was petrified by it. Diana had been meditating for years to help with her anxiety, and she had begged me to join her, but being alone with my thoughts was not my idea of a good time.

Finally, one night during my time as secretary of state, she convinced me to listen to a guided sleep meditation with her. It worked. I felt myself sink into the bed, felt my body go light and my limbs become limp, but then something inside me said, *No!*

Panic swept over me. I felt far too vulnerable, as if someone was about to punch me, and it scared the shit out of me. I'd vowed never to try meditation again, saying simply, "I don't like the way it made me feel."

But now, on a daily basis between therapy sessions, I was lying there, eyes closed, listening to myself retell some of my worst memories. I was accessing parts of them that I'd long since blocked off, for my own safety. This process made me feel as though meditation might be possible. I downloaded the guided meditation app Diana used, and soon it became a daily ritual; sometimes I meditated multiple times per day. I was learning to let my defenses down.

Nick asked me the suicide question every session: "In the past week, have you had any thoughts about killing yourself?" Each time, I said no, and it was true. In fact, right from the start, therapy made me aware that I wasn't going to kill myself, and that alone was a colossal load off my shoulders. Pretty quickly, we worked out that my suicidal ideation had been in a very early phase when I came in. I hadn't done the things that people who are actively considering death by suicide do—I hadn't given away my belongings or begun imagining how I'd commit the act.

"Did it just feel," he asked that first day, "like you'd have been better off dead?"

That was exactly how I felt. He'd hit the nail on the head. I didn't *want* to die. I wanted to live, to be happy, to love my family and do my job. But increasingly those things felt impossible, and I was becoming a gargantuan burden to everyone I loved. I'd told myself I could make people's lives better. Not only had I failed, but I'd made my family's life worse. Logically, it had seemed there was only one thing left to do. And I'd been thinking about it every week.

"That feeling is the first step," he said. "It's a good thing you came in when you did."

Oct 23, 2018

Prolonged Exposure: Second Session

Treatment focused on introducing the concepts of Prolonged Exposure and relating the fallacy of control. Veteran asked appropriate questions and was active in linking experiences from combat to current life feelings and situations. Veteran was asked to observe various avoidance strategies over the week to start building in vivo hierarchy.*

Also, discussed pressure of public spotlight on Mr. Kander and how that could impact therapy. Veteran verbalized plan to keep external interest in his care distant until he completes treatment for himself. Undersigned praised Veteran as this is a good step towards his own recovery.

Right away Nick endorsed my decision to completely shut down any engagement I had with the media, including interviews. It was

---

* The in vivo hierarchy is the list of situations, activities, places, or objects a person with PTSD is avoiding because they perceive them as dangerous.

like the first step in treating alcoholism: remove the person's access to booze. It turned out that Nick was getting a lot of pressure from higher-ups to get me to do videos and interviews about the value and helpfulness of the VA system, and he kept saying no. "This is my *patient*," he'd said. He told me he was worried that they'd go around him to get to me, and that's exactly what happened. When they did so, I flatly refused.

Less than two weeks after my first session, the *Boston Globe* published an article about the impact of my announcement. It quoted an eminent authority on PTSD treatment, who said I had "the potential to become the poster child . . . for post-traumatic growth."

I actually thought it was grossly irresponsible for that guy to say such a thing publicly, and so did Nick. Of all people, he should have known better. I wasn't the sort of poster child he envisioned. If anything, I was the perfect representative of a concept Nick had discussed with me: hypervigilance.

PTSD, it turns out, presents in different ways in different people. Me, I became hypervigilant, experiencing extreme alertness and sensitivity to danger (I couldn't wear headphones outdoors, I couldn't sit in a restaurant without facing the entrance, and so forth). I had also developed behaviors intended to stop danger in its tracks (such as carrying a gun and patrolling the house every night).

"Feel the feelings," Nick said. I was sort of beginning to understand what that meant, but at the same time, I didn't want to feel a lot of the feelings that therapy was dredging up: guilt over almost certainly traumatizing that driver in Kabul, shame over everything I'd put my family through.

Oct 30, 2018
Prolonged Exposure: Third Session
Veteran continues to be motivated and appears some of his intellectualizing and self-defeating beliefs are related to avoidance of emotions surrounding traumatic experiences. Imaginal exposure will help the veteran connect

further with various distressing emotions and ultimately help him reduce his defenses.

Like any graduate program, therapy turned out to require a *lot* of homework. My main assignment was to do in vivo therapy, which simply meant I had to spend forty-five minutes at a time doing things I'd been avoiding for a decade: walk around a park with headphones on (I felt exposed when I couldn't hear); go to a restaurant, sit with my back to the door, and avoid turning around. At first, I couldn't do it. At a restaurant I wound up "going to the bathroom" three or four times so I could scope out the room. But then, slowly, I learned a different way. My brain, a little bit at a time, became able to let go. Nick taught me the vast difference between a behavior I *had* to do and one I *preferred* to do. If, after forty-five minutes of the in vivo assignment, I decided I would rather shift my seat to face the door, that was fine. The point was, I could choose. I didn't feel I had to do it. It sounds like a tiny distinction, but for me, it was night and day. That describes a lot of my therapy—we weren't swinging for the fences; we weren't trying to "cure" my hypervigilance or erase my trauma. The goal was simply to break the grip they had on me—not to make The Monster go away, but to tame it. Domesticate it. Turn it into something that lived in my house, by *my* rules. Not unlike Talia, really.

One day, after the umpteenth rendition of the Sabet story, I told Nick I was getting bored by telling and listening to the same story over and over, which made him laugh with excitement. "Great!" he said. "Boredom is the goal! Let's move on to a different memory." By this point, Nick and I had developed a rapport, and I was beginning to trust him as someone who could help lead me through this process. I still dreaded the sessions—I knew they were going to hurt—but increasingly I was leaving them feeling something akin to hope. Nick always struck the right tone, even joking when appropriate. For instance, when it came time to report on how the homework had gone during the past week, he would rub his hands together villainously. "Ah yes," he'd say, "I love to hear about your suffering." In his self-deprecating way he reminded me that the more my homework sucked, the more likely it was to yield progress.

I now knew I'd made the right call in going to the VA. When I'd first announced I'd be going there for treatment, a number of people were perplexed. "The VA is a nightmare," they said. "You know you can afford to see a non-VA therapist, right?"

Of course I knew how byzantine the VA bureaucracy was— without VCP, I'd have still been waiting for my first appointment. But the question bothered me—it reminded me of the attitude of people who told me in college that I had too much potential to waste it on the army. If anything, my experience with the VA proved exactly why I *did* want to use it: its staff had a thorough understanding of the challenges I was facing, and nothing I said could surprise them. Now that I finally was in the system, everyone I interacted with deeply impressed me.

Perhaps surprisingly, one of the most helpful things at this time was Diana's therapy. Nick made me aware of secondary post-traumatic stress—which I had never heard of—and suggested that Diana see a therapist too.

When you live with someone who constantly tells you how dangerous the world is, how vulnerable you and your family are, someone who wakes you up to recount violent night terrors, you absorb that trauma and make it your own.

Diana had spent a decade sleeping next to a husband who thrashed all night with nightmares, who stalked the house with a gun, who called her every night to ask if the doors were locked and alarms armed, who would bellow with rage when angry, and who convinced her that white supremacists were coming to murder her and her son. This husband's public life had forced her to bear her terror alone, never confiding in anyone, not even family. For years, since I had started verbalizing my depression and my thoughts of being a burden to everyone, she had made sure that whenever she came home with True, she entered the house first. This was a precaution, in case she found me lying dead on the floor with a gun in my teeth. Diana's PTSD was every bit as real and crushing as mine.

For years, as the two of us weathered a hard time, I had been convinced I was married to the most impossible woman alive. Now I understood that it had been my problem all along.

DIANA

Two years before Jason announced that he would not run for mayor, I had gone to a talk therapist to address my growing anxiety, and that experience did nothing. So I found my own cocktail of numbing behaviors: work, exercise, alcohol on the really bad nights, and meditation. It's funny—meditation is often thought of as this amazingly healthy habit, but if you're using it to escape your life and your thoughts, it's an avoidance behavior, just like using drugs and alcohol.

After Jason's withdrawal from the race, a friend recommended a therapist who specialized in trauma and did something called somatic experience therapy. That sounded a little woo-woo to me, but we were in Try Anything territory, so I gave it a shot.

Somatic experience therapy is based on the idea that traumatic experiences can become trapped in the body and wreak havoc on the nervous system until you deal directly with them.

You've heard of the fight-or-flight response. It turns out there's one more option: freeze. Each time Jason would vividly recall a memory or start talking to me about the potential danger we were in, my body would have a physical reaction—my muscles would tense up, my heart would start pounding, my breaths would become short and shallow: I was primed to escape. But with nowhere to go, all of that energy stayed in my body long after the conversation. Without a chance to let my body "reset" in preparation for the next potential threat, I just started accumulating more and more of it until it burst out as anger or manifested as an anxiety attack.

At my first appointment, my therapist, Candy, told me that my own experience had been traumatic. She defined trauma as "anything that is too much, too fast, too soon, or not enough for too long." When she said that, I burst into tears. Hearing those words allowed me—for the first time—to acknowledge what I'd been going through and that it was real.

We spent the first session reviewing every detail of one single anxiety attack that I'd never been able to get out of my mind. A few years before, I'd taken True and his babysitter to the mall in Columbia so he could play at the kids' play area. Because she had recently pointed out that I always seemed to hover over her when she watched True, I planned to wait for them in the food court while I tried to get some work done. As soon as they were out of sight, I began to feel that fight-or-flight response. In my mind, someone had just chloroformed True in the play area and was about to walk out of the mall, carrying his limp body. I began scanning the scene, trying to find the person toting an unconscious child, to remember the clothes he was wearing, but there were too many people moving through the food court. My eyes began to fill with tears because I couldn't keep up with all of them. I began to hyperventilate. I phoned Jason, crying and barely able to catch my breath, and for twenty minutes he stayed on the phone with me, attempting to help me calm down, until he had to go back to work. Even though nothing bad happened that day, my mind kept replaying that vision of someone carrying True out of the mall.

Candy took me through that experience. She helped me get out of freeze mode, where I had been stuck for years, and return to a place of safety, of joy. Like True at the mall that day—he'd actually had a good time.

I walked out of that first appointment feeling like I had just meditated for a week straight. The complete peacefulness lasted for two days. I couldn't wait to go back.

And in the next few sessions, the real work began. I learned that I had welded together my feelings of love and fear. Whenever I felt love for someone, my brain would say, "But what if something horrible happens to them?" And my mind would be flooded with terrifying images.

I learned just how deeply hypervigilance had been ingrained in me. At the beginning of our time together, Candy would say some-

thing like "You know we're safe here. In this room." And for effect, she would gesture around her small office. But I didn't believe her. In fact, I disagreed with her so strongly, just the suggestion would make tears start to flow down my face.

And I learned that scary thoughts don't just happen in your brain—they also happen in your body. Candy taught me not only to notice what was happening to me physically during a moment of anger or anxiety but also to get curious about those feelings and track and describe them, in order to help my body "discharge" the tension.

Diana and I had done everything together since we were seventeen. Why should therapy be any different? At night, we'd compare our lessons and our homework like we had when we were law students together. It was like we'd been chained in a dark cave for a decade, and now our eyes were slowly adjusting to the light.

After years of struggling to understand each other, we suddenly had some tools to use to communicate our feelings. After True went to bed, we'd talk for hours. Sometimes we turned on music and just slow-danced. Diana called it our second marriage. It was wonderful.

Nov 13, 2018
Session Number: 5
Veteran arrived 15 minutes late with a packet of Kleenex in hand. He apologized for being late and stated he came from seeing his grandfather. Veteran disclosed recently learning his grandfather was dying and thus did not complete all homework over the course of the week.

Treatment focused initially on providing support and exploring the Veteran's relationship with his grandfather but also connected this with his current treatment. Veteran was appropriately tearful throughout the session and this was contrasted with previously feeling emotionally numb prior to starting treatment.

His whole life, my grandfather Edward Felix Kander, the one we called Pop, did his duty. When there was war, he went, flying as a radioman all over North Africa, from Casablanca to Karachi. When the war ended, he came home and married Ann Benjamin, the girl who had waited for him. He went to Harvard Business School, but after one semester, he was called home. His father was dying. He never returned to Harvard. His father had left the family poultry business deep in debt. Pop put the business on his back, and for ten years he hated every minute of it, but he rebuilt it, then sold it. He raised three children committed to decency and family.

He lived by a simple rule: "Be an asset to every team." It was impossible to dislike Ed Kander because Ed Kander had decency and respect written on the very chambers of his heart.

I was the first grandchild—the one who named him Pop—and for as long as I'd been alive, he had been there every step of the way, calm, wise, reassuring, a model of duty and devotion to his family. In the weeks before he died, I sat with him every day, along with his kids, his beloved brother John, and the men who had been his best friends since childhood. As much as it hurt to watch him go, I knew how lucky I was: two months earlier, I'd have thought I was too busy to spend so much time with him. Even as he lay dying, I'd have told myself something like "Pop would want me to keep working for this city he loves," but it would have been bullshit, and I'd have known it was bullshit, which would have made me hate myself so much more. When I'm on my deathbed, I don't want to be alone. None of us do, and as selfless as Pop was, he didn't want that either.

He died on November 16, five days after he turned ninety-five. I was so sad to lose him, but I was also grateful. Grateful for the past two months, sitting around reading the paper with him, or watching ball games, then going to eat lunch with him at the café in his retirement home—he'd insisted on moving there, so as not to be a burden. Then he'd race True down the long hallways, with True sprinting to keep up with Pop's motorized wheelchair.

I distinctly remember sitting in Nick's office and feeling grateful for something else: those past few weeks of therapy had already given me some of the tools I needed to access that sense of gratitude. I still had a long way to go, but I could feel a thaw.

"Feel the feelings," Nick said, and I tried, and slowly they began to come.

I felt terrible about losing Pop, but at least I *felt*. The grief wasn't dampened or distanced; it was real and it hurt, and for once, my brain didn't smother it. It let it breathe. It let it hurt. I was slowly regaining the ability to feel things, whether good or bad, and I was aware of it. It was like getting taller and—day to day—noticing my feet extending farther off the end of my bed.

In therapy I could make changes only if I understood how things had gotten so bad, and with Nick's help, I began to see where my emotional numbness originated. It was a coping mechanism. For years, I had been inundated with negative emotions—sometimes as a result of traumatic memories and sometimes as a result of shame caused by that trauma. Either way, it was disturbing, so my brain deployed countermeasures to suppress that disturbance. As my mind pushed the emotions deep down inside, I helped out by finding ways to avoid them. This was why I stayed busy, and when I couldn't stay busy, why I had to listen to music or a podcast or the MLB Network* or some combination thereof. Anything to keep me from being alone with my own thoughts.

There was nothing inherently wrong with this natural defensive reaction, except one thing: it wasn't a laser-guided smart bomb but rather a big messy explosive that destroyed everything within its kill radius. My brain couldn't tell the difference between negative emotions and positive emotions, so it suppressed all of them. The result: I grew distant from Diana, from my son, and from my parents and brothers and oldest friends. The negative emotions had become so unbearable, I'd crushed them, and in so doing, I'd crushed many things that meant the most to me.

I'd stuffed joy and love and sadness and disappointment in a card-

---

* I hate to think of my Royals as a coping mechanism, but sports has many wonderful purposes, and one of them is to provide some entertaining escapism.

board box marked FEELINGS, thrown a tarp over it, and shoved it into a dark corner of my emotional basement. And now, slowly, I was rooting around in that basement and finding some stuff I could still use.

Nov 27, 2018
Session Number: 7
Veteran described having difficulty on recent trip sitting in a running car (not moving). He noted high SUDs, heart racing, and this as discussed as focus for in vivo this week. He also wants to watch *Zero Dark Thirty* as in vivo and thinks he may enjoy that now, because he has been able to read his combat book with no nightmares and no other re-experiencing symptoms.

For years I thought my nightmares were brought on by the media I consumed or the places I'd gone during the day. I had these horrific dreams every night, so I was always trying to figure out what caused them, and I could always manufacture a plausible trigger: *I heard that NPR story about Afghanistan*, I'd tell myself, or *I saw that trailer for a war movie*, or *I'd been in that crowded restaurant with my back to the door.*

I was always thinking about the war. It seemed that everything made me think about it, so I was always able to point to something and seize on it as a trigger. This is why I'd stopped watching war movies (or any movies involving a kidnapping) and why I'd given up reading books about the military. I thought of them as junk food, in that they tasted good going down but wreaked havoc on my system. My evidence: while I watched or read these things, my heart raced like it did when Nick and I did our prolonged exposure sessions.

So years ago, I made all these changes and then, when my nightmares stopped taking place in Afghanistan, I told myself that my efforts were working.* But I was still having nightmares.

---

* I literally wrote as much in my previous book.

Nick helped me see that I'd been doing things exactly backward. Apparently, this was called "avoidance," and rather than helping, it was hurting. My nightmares weren't the result of my thinking about Afghanistan but rather the result of my trying so hard *not* to think about Afghanistan. I spent my days playing Whac-A-Mole with intrusive thoughts and disruptive emotions, but at night, when I was unconscious and my guard was down, they came rushing back in.

"Your brain needs to process these things," Nick said, "and it's trying, but you're not letting it."

"So if I process it while I'm awake," I said, "I won't have to while I'm asleep?"

"That's the idea."

Sitting there in Nick's office, I had an insight. Since coming home from Afghanistan, the only times I made it through the night without terrible nightmares were when I was on army duty.

I told him this and added: "That never made sense to me either, because sleep deprivation had always made my nightmares worse, and I had some late nights and early mornings at Fort Leonard Wood. But I often slept great."

"That makes sense though, doesn't it?" he asked. "I mean, the nightmares come because you're suppressing those memories, but it's hard to imagine how you could suppress them while you're training people for combat."

So part of my homework became watching movies and reading books I'd been avoiding for the past eleven years. I felt some trepidation about it, so I did it a little at a time.

I started with *Once an Eagle*, by Anton Myrer, which was assigned reading in the Hoya Battalion and, I believe, at West Point as well. Authored by a marine who fought in the Pacific during World War II, it's arguably the best novel about combat (and about leadership). I used to reread it all the time, but a few years after I came home, I'd stopped.

Sure enough, Nick had been right. My SUDs would spike during and after watching or reading such a work, but then they'd come down, and my sleep was improving. Sometimes when I read or viewed military content, the SUDs didn't spike at all.

I loved the military, and I had missed the military, and now I embraced the fact that I could delve back into that literary and cinematic world. I had a decade's worth of stuff to binge on. Lord knows I had the time.

Zero Dark Thirty, about the operation to find and dispatch Osama bin Laden, was at the top of my list. After all, 9/11 had inspired me to go to Afghanistan in the first place.

> Dec 11, 2018
> *Session Number: 9*
> Is this the final session of this Evidence Based Psychotherapy?
>> Yes
> Veteran discussed completed in vivo homework over the past week and discussed very little distress related to any of the activities.
>> Treatment focused on planning to use more CPT/guilt related elements in treatment and to use Krakow nightmare rehearsal therapy.*

After eight weeks, Nick considered me stabilized. I disagreed.

Between working out, meditating, reading for pleasure, and doing my therapy homework, I was piecing together a daily routine, but my days and weeks were still these barren wastes of emptiness and fog. I felt I had taken only the first steps on the path to becoming stable.

But I could see progress too, especially in how I was able to pour my attention into True and Diana. I started to learn how to cook, or at least how to prepare nonlethal food. I acted as a personal

---

* As part of my homework, Nick had me write out one of my typical recurring nightmares, but with a different ending: instead of being kidnapped or killed, I escaped. The idea was to train my subconscious to incorporate this as an actual possible ending. Sort of like the plot to *Inception*, but in reverse.

assistant—albeit a terrible one—to Diana in her consulting and speaking business. That didn't last long, but it did help fill my day and give me opportunities to feel useful.

And I became a full-time stay-at-home dad. I dropped True off at school every morning and I picked him up every afternoon. We used LEGOs to build things together, played video games, did puzzles, drew pictures. We would spend whole afternoons at my parents' house or playing with his cousins. We loved watching Royals baseball games together, and I taught him about all my favorite players when I was growing up. He had his own favorites from the current team. And when I was with him, increasingly I was actually *with* him. I wasn't checking my phone, I wasn't thinking of my next trip. I was with my son, and there was nowhere else I wanted to be.

These moments of joy were laced with anxiety. I couldn't stop imagining what it would be like for True if I couldn't fully recover— what if he finally got his father back, only to watch him fade away again? Every time I caught myself wanting to argue with Diana or breaking into a sweat because some guy was tailgating me, fear raced through me, the kind you have when you're a kid in bed at night and something goes bump in your closet. The Monster was still there. If I let my guard down, it would roar back, stronger than before. This was a problem because letting my guard down was exactly what I was supposed to be working on.

Understandably, I was in no hurry to stop therapy. I felt like someone who was learning to walk again with the help of crutches and a battalion of physical therapists, but I was terrified that if ever those crutches and helpers went away, I'd collapse again. As we neared the end of our prolonged exposure course, Nick started making noises about ending our sessions, and I got very agitated. This was working, and nothing else had. Why would I want to stop?

Dec 19, 2018
Individual Psychotherapy Progress Note
Focus was related to cognitive distortions around fluctu-ating mood. Veteran tended to catastrophize that he had

not actually been benefiting from treatment, maybe didn't have PTSD at all. This was related to 2–3 days of a lower mood.

I truly did not understand that therapy was not a linear thing. You don't just get a little better every day. It's more like you have some good days, and you're almost scared it will all evaporate, and then one morning you wake up and all the darkness roars back in. In December, three months in, I was in a deeply bad place for three straight days. It didn't make sense—I'd thought I was getting better, and now suddenly the nightmares—which had become a biweekly event on average—were once again torturing me all night, every night. The profound shame and dislike of myself came back too.

I dragged myself out to a sushi restaurant with Diana and True, and we let True watch his iPad with headphones on so we could talk. I remember putting my head in my hands and asking her, "Do you think I'll ever be happy?" I was terrified that the symptoms I had when I was first diagnosed would be with me forever. That the depression was permanent.

But when I talked to Nick, he almost smiled. "Three days?" he said. "Don't worry. It's not depression unless it lasts for two weeks. You're just having a shitty week."

It sounds banal, but that was a revelation. I didn't *know* I was allowed to have shitty weeks. I had never before given myself the space to have even one shitty day.

That day, before I left his office, Nick said it again: "Feel the feelings."

"If you feel shitty, let yourself feel shitty. Don't fight it," he said. "And if you feel good, feel good; don't shame yourself for feeling good. Just feel the feelings."

In that moment it was as if the gray housing of cliché fell away from that saying, and I could see the elegant emotional machinery whirring away underneath. Suddenly it made perfect sense.

Recovery is full of these clichés, like "One day at a time" in addiction therapy. They sound silly from the outside, but now I understood why they hold power. They are like mantras—simple, clear, true concepts that I could fall back on when I didn't know what I was supposed to do. For years, confusion had been my mortal enemy because confusion on a battlefield, or in a meeting with a narco-warlord, can be lethal. So whenever I encountered anything that confused me, my brain smashed down the anger or self-hatred button because those emotions made sense. I knew what to do with them.

Now I realized that over the past three months, whenever I'd just let myself "feel the feelings," no one had gotten hurt. Nobody had crashed through the door with assault rifles and black hoods. But what *had* happened, in fits and starts, was something very much like joy.

> Jan 03, 2019
> *Individual Psychotherapy Progress Note*
> Focus was more toward existential thoughts/crises that are arising related to work and future. Veteran discussed feeling "plateaued" in treatment and the undersigned confirmed this is typical toward the end.

As 2019 began, I had been a complete ghost for three months. I hadn't given a single speech. I hadn't done any events. I hadn't even tweeted, except for some language Let America Vote gave me on election day in 2018. I had vanished completely. For people who live in the public eye, that kind of hiatus can become permanent in a hurry. By February, I was watching from the sidelines as people I knew began to announce their candidacies for 2020—Pete, Kamala, Elizabeth, Joe, Cory, Beto. I spoke to a lot of them before or after they announced, and when I did, I had to ask myself, *Did they have as good a chance of beating Trump as I once had?* I wasn't jealous—by now I knew what would have happened if I'd gone through with the run—but I still felt dread and anxiety about whether anyone else could win.

It was bizarre, really—I was just this stay-at-home dad who happened to have long conversations once a week, sometimes while walking through Costco, with presidential candidates who were asking me for advice.

I'd also begun to sweat over my career, which for all the world appeared to be dead. I simply had no idea what the hell I was supposed to be doing. It was very strange to know that, inside my private bubble, I was progressing, but as far as the outside world was concerned, I was the smoking crater where Jason Kander had fallen back to earth. I *had* been that, once, but now that I was climbing out of the hole, it was confusing to have so much pity projected onto me while I thought I was getting better. There was only one thing I knew for sure: I wasn't able to go right back to running for office. I didn't even know whether I wanted to, but I didn't get the sense that anyone else wanted me to either. That realization gutted me. Aside from the military, I'd never found anything else I was good at.

As pathetic as it sounds, the idea that I'd lost access to the cool kids' table deeply bothered me. I didn't like this about myself, but I worried I'd been interesting to celebrities only as long as I'd been valuable to them. Now that I was just another guy who used to be a somebody, a guy who took out the trash and could barely fry an egg, I figured they were over me. This was tacky—I had so much going for me at home, I shouldn't have cared. But I wasn't a saint. I did care.

> Jan 09, 2019
> *Individual Psychotherapy Progress Note*
> Treatment focused on guilt/shame. It was determined
> that Mr. Kander has more shame related to what he was
> not able to accomplish in the service vs. guilt. We looked
> at the role beliefs of control over a traumatic event have
> in getting stuck in recovery.

Afghanistan had taught me one lesson above all: that I would be safe if—and only if—I could control a situation. So my brain registered any discomfort, anything I couldn't control, as a life-or-death threat. My hypervigilance, my self-loathing, my workaholism, my anger: they were all about control.

In one of our early sessions, Nick asked me what percentage of my daily life I believed I could control. My answer was 85 percent.

He said that the average person will answer somewhere around 50 percent, while trauma survivors often choose a much higher number, like 80–90 percent. The truth, he said, is actually about 3 percent.

The need for control in every part of my life felt like a matter of life and death. When I didn't have control, I felt enormous levels of stress, as if I was struggling to survive.

I often turned to anger or shame so that I could feel some sense of control. It worked like this: If I felt anger at myself, I had something to focus on, and that focus *felt* like control. Likewise, if I loathed myself and decided I was simply a bad person, I felt I knew something for sure. In a sense, this craving for control had enabled me to work so hard and rise so fast. But in practice, mostly it put up brick walls between me and the people I loved.

To paraphrase Yoda's wisdom about the path to the dark side: fear of losing control leads to anger, and anger leads us to cause others to suffer.

Today, the statement "I can only control 3%" is written on the wall in our basement workout space.

One of the most important skills I learned in therapy is one I'm still working to master: becoming comfortable with being uncomfortable. And for me, that's largely about accepting the fact that 97 percent of what happens is beyond my control.

DIANA

Anger never travels alone. It's like a bouncer, protecting the fears and vulnerabilities we're afraid to voice. Anger makes us feel like we're in control of a situation, like it's protecting us. But in reality, it's preventing us from having the real conversation that could help.

It's taken a lot of work and practice, but Jason and I rarely get angry at each other anymore. Instead of zooming out of the particular issue and bringing in past disagreements and wrongs in order to win the fight, we zoom in and get curious about the specific issue we're dealing with and what might be behind the anger. One of us might say, "Let's talk about what we're each trying to control and maybe the fear that exists behind it."

When "you can only control 3%" became a mantra in our house, we even wanted to get matching "3%" tattoos. That is until we found out that the Three Percenters are a far-right anti-government militia. In 2018, they were considered the most dangerous extremist group in Canada and have been responsible for substantial violence in the United States. So we're still workshopping the tattoo ideas.

The need for control had—long ago now—linked up with another need: for survival. Together, the two had spent years becoming the Bash Brothers of my brain.

I know a couple of things most Americans—thankfully—will never learn:

1. What it feels like to believe you are about to be killed violently
2. What it feels like to prepare your mind to violently kill another human being

As a result, my brain has a tendency to treat every threat—whether to my person, people I care about, or even my career—as though it is the greatest possible threat: death. It's like those carnival strength tests where you swing the sledgehammer to see how high you can make the ball go. If it hits the top, the bell goes *ding*. Normal, untraumatized brains can triage threats—*this one is minor, this one is extreme*—and react accordingly, with appropriate restraint. I came home with a brain that found it difficult to do that; it was a sledgehammer, and every single threat, however minor, would ring the bell at the top of the threat meter. That's why, when I lacked control over events (like elections, for instance) and had to wait for them to unfold, I felt like I was dying.

Suddenly I understood what had happened on those election nights—why a sense of impending doom was the only feeling that welled up.

Now, when my brain tries to send me that kind of signal, I at least know that it's not me, it's The Monster, and The Monster is a horribly unreliable adviser. I explained this to a friend once, and he said, "As GI Joe taught us, knowing is half the battle."

Another thing—I've learned that my brain wants to return to the simpler operational environment of combat. To help me understand why I got stuck in my experience of that world, wartime Afghanistan, after only four months, Nick went to his whiteboard. He drew a line down the middle and wrote "Here" (home) on the left side and "There" (Afghanistan) on the right.

On the left side, he asked me to list the principles and beliefs by which I tried to live my life, the ideas that motivated me and guided my behavior. "But there's a catch," he said. "It has to be stuff you believed in before you deployed and still aspire to believe in now."

"So, like, what my parents taught me?" I asked.

"Sure, let's use that," he replied. "Just give me everything you learned from your parents and used in your life before you deployed and that you still seek to use now."

He held a black marker in his hand, and as I talked, he wrote:

Work hard
Hustle
Help people
Stand up for people
Be humble
I can do anything
Obligation to others
I am privileged
Don't steal
Don't do drugs
Don't cheat
Don't start fights

Do finish fights
Fight for others
Don't kill
Be faithful
Be protective
Care about the world
Have high standards

When we got to the bottom of the whiteboard and ran out of space, we stopped, and he said, "Okay, now let's do the same thing for Afghanistan. What principles did you learn over there that you used all the time?"

Be calm
Suck it up
Someone has it worse
I am lucky
Danger everywhere
Always have weapon
Protect self and others
Maintain control of situation
Eliminate risk
Eliminate threats

Only about a third of that side of the whiteboard had been filled when I stopped. "That's it. I can't think of any other ideas."

He pointed to the list on the left, the longer one, and asked me about the point of those principles—what did they help me do? At first, I struggled to answer the question, so he framed it another way. "Why did your parents teach you those things, as in, what do you think they wanted for you?"

I thought about it for a moment and then, with a shrug of my shoulders, said, "I guess they just wanted me to make a difference."

"Why?" he asked.

"Because that's what made them happy, and they thought it would make me happy too."

"Yes! Good!" Nick liked that answer.

"Now, this other list, what the army, or really, what Afghanistan taught you. What was the point of this? What did it help you do?"

I didn't have to think to come up with that answer. It was easy. "Not die," I said.

Nick said, "Bingo," as he uncapped a red marker. He wrote "Survive" at the bottom of the Afghanistan list. He went on to explain that my brain got stuck in that world and in fact didn't believe that I had really *left* it.

He drew a line between the two lists with the red marker and put an arrow on each end, as if to indicate they were trading places. "What would happen if you went to Afghanistan and operated only by the principles your parents taught you?"

The answer was obvious. I made a *pfff* sound and rolled my eyes, and Nick laughed. He wrote "Death" over the two-way arrow.

Then he drew an identical two-way arrow beneath it and asked, "What would happen if you operated at home the way you operated in Afghanistan?"

"I think I kind of have been," I said.

He explained that I still had all the values my parents taught me, and I was still using them, but now these others were involved, and a part of my brain wanted to use *only* the values I'd learned during my deployment. It was worried that if I didn't, I'd die.

"The Monster kept you alive over there," Nick said, "but over here, it's unhelpful."

Then, with the red marker, he circled one of the items on the Afghanistan list: "Someone has it worse."

"We've talked about how this idea kept you from acknowledging your PTSD," he said, "but really, it just boils down to this . . ."

He switched to a new marker, a blue one, and wrote "Guilt/shame."

"You brought this idea, that you didn't do enough, back with you too, and it made you think that the stuff your parents taught

you—your core values—actually came from your guilt and shame."
He pointed again to "Guilt/shame." "The Monster made you believe you were trying to make a difference in the world only because
these feelings were hanging over you."

"But that part of me—the part that wanted to make a difference—
was already there before I deployed," I said.

"Exactly, it was already there," he said with a nod.

Then Nick pointed out that the list on the right was shorter because survival is objectively easier to understand than making a difference and achieving happiness. "Our brain likes simple," he said.

"That's partly why I wanted so badly to go to Kandahar," I said,
almost dazed with new understanding.

Sagely, Nick nodded. I could see him smile.

Months into therapy, I felt like I was finally entering recovery. When
symptoms like anger and hypervigilance did rear up, I could recognize what they were about. I could put a name to them, and more
than that, I could see their cause. I could sit in a restaurant longer
with Diana and True, sometimes even after the meal was over, instead of feeling uncomfortable staying in one place for too long.
I was able to walk the dog while listening to a podcast instead of
feeling the need to access all my senses in order to detect danger.
And increasingly, when something upset me, I would make myself
talk about it with Diana instead of numbing myself and avoiding it.
More and more, I learned to be comfortable with being uncomfortable.

Then one day I was folding laundry with Diana and I said casually, "I really didn't have a great day today. But it's okay. I think I'll
do better tomorrow." She stopped what she was doing and looked at
me, teary-eyed, then walked over, hugged me, and buried her head
in my chest.

Quietly, she said, "I haven't heard you say anything like that in so
long."

11

THE STUFF I WISH SOMEONE
HAD TOLD ME

Spring drew near and I was getting better, but gradually a new insidious thought crept in: I was bothered by how much better I was getting. How was it possible, I asked Nick, that so many people suffered from PTSD for years, yet I felt almost like a new person after just a few months? Did this mean that I hadn't really had PTSD?

See, The Monster works like a nightmare houseguest: it moves into your brain, trashes the place, and then gaslights you into believing it was never there.

PTSD convinces you that you never had it, and you're just an asshole. And I kind of felt like one. I assumed I was having a completely different experience of it than anyone else did. All across the country, vets—and countless other people—were trying to fight mental illness without a fraction of the advantages I knew I had. For instance, one of the first questions the VA asks when you begin treatment is "Do you have a support network?" And I was able to say that I didn't just have extraordinary friends and family, but also a national support system of people who wanted to help. I had financial advantages too—Diana and I weren't superrich, but we had money, and more than that, we had deliberately saved up two years' worth of income because we wouldn't be able to earn money when I ran for president. (This was back in the innocent, dewy-eyed days of 2018, when things like conflict of interest still mattered.)

When I brought up these concerns, Nick was ready. He showed me studies done by the VA's PTSD researchers. As it turned out, more than half of the VA's PTSD patients get better. I hadn't known that—and I realized that the reason I hadn't known it was because I'd never seen it. In films or on TV, victims of PTSD might be depicted with harrowing accuracy, but only a limited range of them. They were the ones beating their spouses or robbing banks or abusing alcohol or doing all three at the same time. I couldn't think of a single example of a character who had recovered without performing some monumental act of redemptive heroism.

Before treatment, I had believed PTSD was terminal, that it would end my career and possibly my life no matter what I did. I dreaded the diagnosis more than the symptoms, which I believed would go away if I could just redeem myself by achieving this or that. Redemption seemed like a mirage. Recovery, on the other hand, is a very real possibility—even a likely one.

The odds, it turned out, were on our side. Nick showed me that the greatest predictor of whether people got better was whether they committed to the program—whether they did the homework.

"You're supposed to get better," he said. "You did the homework. *You* tamed The Monster."

I decided I believed him, but to this day The Monster still knows how to raise doubt. *Don't blame the war,* it says. *You had a stressful career and you couldn't take it, that's all.*

When those thoughts creep in, I remind myself of two things:

First, if that were true, why should I care? I used to be miserable, and now I'm not. It doesn't matter why.

But since that's not enough, and because my ego struggles with the idea that The Monster never existed and the problem was all just me, I bring the nightmares to mind. For more than a decade I'd been terrified of what would happen when I fell asleep. These days, I get nightmares occasionally, but not with the same severity, and they disrupt my sleep only about once every two weeks. So when The

Monster tempts me to question the legitimacy of my diagnosis, I just think about the nightmares, which reminds me that The Monster is real—I know because I fucking tamed it.

And so, about six months into weekly therapy, I considered the possibility that I was healing. That's when Nick told me he thought I was ready to appear on TV again. He'd known all along that one of my goals was to get healthy enough to say yes to media requests. It wasn't—at least I hoped it wasn't—because I missed the attention. I wanted to demonstrate post-traumatic *growth*. When I was diagnosed, I had other vets like Stephen and the VCP guys to look up to, but there weren't really any public figures appearing on screen who had fought this beast and won. I wanted to be that figure.

But when Nick brought it up, I balked. Fame had been a substance I'd abused, and the media had been my dealer. What if my growth in therapy had been partly a matter of "getting clean"? Was it smart to tempt that addiction—was it quicksand I'd never be able to climb back out of?

"If I had come here with alcoholism," I told Nick, "and you'd helped me get sober, you wouldn't try to get me back to my job as a beer taster."

Nick seemed to find that metaphor persuasive, and so he didn't bring up the matter for a few months. Then, during one appointment when I was feeling particularly good, he asked me whether I had any current goals for the future. By this point, I'd been working out for months and was surpassing my previous peak level of fitness, which I'd hit at army intelligence training in my mid-twenties. I shrugged and said, "Eh. I think sometime this summer I'll probably have six-pack abs. Never done that before."

Nick laughed, which made me laugh too. "Not long ago," he said, "your personal goal was the presidency, and one of the things we worked on was temporarily shrinking your world."

"Well, I've done that," I said with a smile, "and it has obviously helped."

"That's good, but your world can't stay this small forever. What if there's a middle ground?"

He leaned back in his chair, and I could tell he was putting thought into how best to communicate something important.

"I know we've been going with this alcoholic beer-taster analogy, but I've been thinking more about that, and what if you weren't an alcoholic after all? What if you were just a person who liked a drink every so often, and now that we've made so much progress on your underlying trauma, you can have a drink or two and stop?"

In June, ten months after my announcement, Diana and I flew to New York to record an interview with Lester Holt for *NBC Nightly News*. I was nervous—not about the interview, but about how much I might *enjoy* the interview. I woke up that morning to a long, supportive pep-talk text message from Elizabeth Warren. Yet another reminder of how blessed I had been in the support-network department and how much of an opportunity my platform provided to make a difference in the fight against stigma.

The network sent a car to bring Diana, True, and me to 30 Rock, and on the ride there I felt worried. What if the need for external validation and public attention came back?

The process felt like slipping into a beloved old pair of jeans. I sat in the makeup chair, held still in my seat while they framed the shot, tried to drink enough water to avoid dry mouth but not so much that I'd have to pee, and waited for the rush of adrenaline that sets in just as the interviewer winds up the first question. I'd done this drill many times before, but something was different. The adrenaline rush never materialized. I enjoyed talking to Lester, but I didn't experience the nervous energy that came with aspiring to do more than simply answer the questions. In the past, there'd been more at stake: a need to expand my audience, make a good impression, win votes, raise money, or appear presidential. This time, the stakes were simple: tell my story, try to articulate what it feels like to have PTSD, and persuade viewers, especially those who needed to hear it most, that post-traumatic growth was worth pursuing.

Lester's questions were much more personal and emotional than any I'd ever answered publicly, and yet it was the easiest interview of my life. I didn't reject the premise of a question, I didn't give dodgy answers or pivot to talking points. I had no talking points.

When it was over, Lester shook my hand and we chatted for a few minutes about his own experiences covering traumatic events. As I was disconnecting my mic and removing makeup, a member of the crew came up and confided his own mental health challenges. Diana and I retrieved True from the next room, where he had been entertaining a team of junior producers, and then took some pictures of him behind Lester's anchor desk. Then we at last fulfilled our promise to visit the LEGO store with him downstairs. As I tried unsuccessfully to persuade True to choose the Statue of Liberty set over the latest *Ninjago* dragon, I realized something beautiful. My mind wasn't replaying the hour prior or planning out my tweets for that evening when the interview was broadcasted. It was on True. It was on showing him New York. I was present.

It was pouring rain, so we weren't picky about lunch and ducked into the nearest sports bar. As I squeezed a lemon into my water and True colored on the back of a paper place mat, Diana asked, "Okay, it's done, how do you feel?"

It was *the* question of the day, so I paused a moment to fully evaluate—and to feel—the feelings.

"I feel good," I said. "But I don't feel *high*."

It was a relief to simply be pleased about what the interview could mean to people with PTSD. I didn't need to do it again.

That night, while my first public comments in ten months were airing in prime time on national television, we were at a pizza place with friends.

Today, The Monster isn't gone. I really struggle when I'm in a very busy public place with my family. If I'm at an amusement park, say, it's still hard to be present rather than "on patrol." But I've gotten a master's degree in myself now, and here's the stuff I wish I'd known when I came home.

Either you deal with your trauma, or your trauma deals with you.

Perhaps my most important pre-therapy realization was included in my letter announcing my intention to drop out of public life to seek treatment: "So after eleven years of trying to outrun [post-traumatic stress disorder], I have finally concluded that it's faster than me. That I have to stop running, turn around, and confront it."

If it were possible to outrun PTSD, I would have; nobody was running away from it faster than me. But it caught up to me, and it was always going to. The secret: I didn't have to run away.

Trauma is like credit card debt. The payment is bad enough, but the interest will ruin you. My initial trauma wasn't as terrible as what some other people have suffered—I wasn't shot or blown up or sexually assaulted. But I *fixated* on that comparison, constantly downplaying both my trauma and what it did to me. (I still sometimes do this—can't help it!) So I let the pain grow and fester until it was completely debilitating.

PTSD is an injury. That's all it is.

Imagine if, instead of getting knee surgery and physical therapy all those years ago, I had just ignored my injury and waited for it to get better on its own. Imagine if I had said, "I'll walk it off," and kept running and playing sports, thinking that eventually my knee would be *forced* to heal. Obviously, it would have gotten much worse. My whole leg would have been mangled, and healing would have required much more medical attention, time, and effort.

That's what happens when you don't treat an injury, and trauma is no different. It's not wine. It doesn't age well. It's more like an avocado. Nobody builds "avocado cellars."

It's not a contest.

I spent years unfavorably comparing my combat experience to that of others. I talked myself into the idea that what I'd done didn't count, that I couldn't have PTSD because I didn't "earn it." As if I could dispatch my trauma by belittling it.

What a giant waste of time that was. One person's brain doesn't know what another person's brain has experienced. That's why I never shame someone else for their trauma, and why you shouldn't either. I'm blessed that many people feel comfortable sharing with me their traumatic experiences or mental health struggles, whether it be losing a loved one or surviving cancer or a car accident. But I've noticed they frequently begin with an apology or a disclaimer, like "I wasn't in a war or anything."

When they do that, I stop them and say, "It doesn't matter because your brain doesn't know what my brain went through. It knows only what you experienced. What happened to you is all that matters in this conversation, and what happened to me is totally irrelevant."

Thinking "other people have it worse" doesn't actually diminish your own trauma, it just diminishes your power to heal.

Mental health is physical health and physical health is mental health.

I'd thought I had "lower back problems," but much of my discomfort was caused by stress. After my announcement, my back pain became so severe that my first several therapy sessions took place while I lay on an ice pack. Now that I'm mentally healthier, I have far less pain and can exercise with a frequency I hadn't been capable of since I was in the army.

I went almost twelve years without a good night's rest. *Some people just don't sleep well*, I thought. Now that I do sleep well, I know the truth. Sleep problems can be a "check engine" light. Your brain might be signaling that something is wrong.

And, by the way, now that I can sleep so much better, it feels like a damn cheat code.

Treat yourself as you would a good friend.

I still don't know how much of my political career was about fighting battles to help others and how much of it was about fighting battles to convince myself that I wasn't a piece of shit. I may never know, and I still struggle with this. But I try to remember that it doesn't actually matter.

It took Nick a long time to convince me to stop thinking that way. One day, he walked over to his whiteboard and wrote, in giant letters, "LOSE/LOSE."

He explained that I had found a way to rob myself of a sense of accomplishment: if I didn't do enough, it was proof that I was worthless, and if I did do enough, I'd managed to do it only because I *believed* I was worthless.

Survivor's guilt is a part of this too. When I was inclined to feel good about something, I'd rob myself of that good feeling because T.J. couldn't see and Kevin was dead. What right did I have to enjoy my family or take pride in a professional achievement?

When I shared this with Nick, he asked, "What would you say to Stephen if he said that to you?"

Of course my response to Stephen would be a lot more kind and compassionate. I'd tell him how this empty gesture hurt him and didn't help anyone. Nick helped me see that I could offer myself that same level of kindness and compassion.

I still struggle with this, but I try to remember to treat myself like a friend and cut myself some slack.

"Letting go of shame," Nick told me, "is not the same as absolution. You're not rewarding yourself; you're just deciding not to punish yourself so much."

Occasionally, instead of using Stephen as an example, he would ask me what I would say to True if he'd said what I'd just said. And I'd reply, "No one is all good or all bad, but you do your best, and I love you." I'm still learning to offer that level of advice and compassion to myself.

In that vein, I'm much better now at giving myself credit for my pre-therapy accomplishments. Yes, the pace at which I worked was at least in part a response to trauma, but the causes I chose were a reflection of who I was—who my parents raised me to be. The Monster didn't make me care about veterans' health care, or ethics reform, or voting rights. That was all me.

There will always be new challenges and possibly new traumas.

In August 2021, as American forces undertook the historic evacu-

ation effort in Kabul, the Taliban retook Afghanistan. To say I found
the images on television triggering would be an understatement.

Intellectually, I'd long expected the Taliban to eventually regain
control of the country, but emotionally, I was unprepared. It was
hard not to feel pangs of doubt—had any of the sacrifice been meaningful? Every talking head on TV seemed quick to write off Afghanistan as another Vietnam or lump it in with Iraq, as if looking back
on a single combined war rather than two distinct conflicts.

I booked an appointment with Nick and showed up at his office
ready to talk through my need to control an uncontrollable situation. "You're right that most of this is about control," he said. "And
just about every Afghan vet I've treated came in to see me this week,
and each one has expressed similar frustrations with the way this is
being talked about publicly." And then Nick gave me some unexpected advice. "But unlike any of them," he said, "you can actually
affect how people talk about this, and maybe you should."

I had been turning down interview requests because I thought
I was too emotional to go on television. But here was my therapist reminding me that America might need to see some of that
emotion—or, more important, my fellow Afghan vets needed to see
America seeing some of that emotion. So I spent a week saying yes
to every national media inquiry that came in. In these interviews
I made the point that Afghanistan was *not* Iraq or Vietnam, and I
reminded Afghan vets that parts of our mission had been accomplished. We who served in Afghanistan could feel both proud and
sad at the same time. I received thanks and encouragement from
hundreds of fellow veterans, and that made me feel useful.

Increasingly, however, I started hearing from people who were
trying to get Afghans they served with into the airport in Kabul to
save them from the Taliban's retribution. I knew how fortunate I
was that my own translator, Salam, was safe in the United States,
and I felt deeply for anyone who still had someone in Afghanistan.
Then I spoke to Salam and found out he still had family there—
they had also assisted American forces and were now being hunted
by the Taliban. I made contact with Rahim, Salam's cousin and the

leader of Salam's twelve family members remaining in Afghanistan. I reached out to fellow Afghan vets—some I'd served with and some I hadn't—and together, communicating by cell phone, we attempted to shepherd people into the airport. I expected this effort to last for a few nights—staying up, living on Kabul time. We worked with the marines on the ground and got some of the families close to the gate. The marines were carrying pictures of our people and trying to spot them in the crowd while we coached our Afghan families on how to use prearranged signals in order to be pulled inside.

Then the bomb went off. The marines—and a lot of Afghans—were killed. It was shocking. Yes, I'd ordered soldiers into harm's way before, but never mothers with their babies. But by some miracle, the women and children in our group had been just outside the blast radius.

Within days, the airport was closed. And on August 31, the US military left Afghanistan for good. My friends and I hadn't gotten a single person out, and now it was hopeless.

Salam's cousin Rahim and I had become close, and when he called me, his mother and his wife and his little girls were all crying. Against my better judgment, I promised him I'd find a way for his family to escape, though I had no idea how that might be possible.

"You are my brother now," he said.

My friends and I started planning together, networking like crazy with former military and intelligence community colleagues—some still serving, some not. This new mission quickly became all-consuming. I kept living on Kabul time, sleeping on a mattress on the floor of my office. The operations tempo felt like another deployment, but worse—not being there on the ground made us feel powerless. One guy I was working with said the situation had helped him understand how drone pilots could develop PTSD without ever leaving the United States.

Everything we tried seemed to fail, and the Taliban was consolidating control. It seemed only a matter of time before the people we were helping were found and never heard from again. At first, I'd focused on Rahim's family: I just wanted to get twelve people out. But before long, I had become a de facto leader of our little group, and now we were trying to save dozens and dozens of families. They were counting on me and a few of my friends to deliver them from hell.

I could think of nothing else. At dinner with Diana and True, I'd be a zombie—my entire mind was elsewhere. I was irritable, hyper-vigilant, and short-tempered. I was never more than an arm's length away from my phone, and the ringer was always on. Diana became understandably worried about me, and one night at dinner True said, "This feels like before."

And that's when I realized that this experience hadn't been just a trigger. This was a brand-spanking-new trauma. I went to see Nick, and he calmly concurred, then reassured me. I had the tools to navigate it. He was right: something *was* different now. I *did* have an idea of what to do.

Instead of going numb with anger or withdrawing, Diana and I started talking about what we were feeling. Like detectives, we became curious about our emotions. We hugged a lot. We used the lessons learned over the previous two years to get through this situation together. Neither of us tried to go it alone.

Salam told me he appreciated all I was trying to do but urged me not to chase the impossible—to take care of myself. But I didn't stop trying to save Rahim and his girls, or any of the families whose pictures and vital documents now lived in my phone.

I knew this was going to take more than just military friends, so I called Kellyn Sloan, who was now at the Bonner Group, one of the top political fundraising firms in the country. I told Kellyn we were going to need a lot of money, and—without ever charging a penny—her team started raising it.

In late September, Operation Bella—named for Diana's refugee grandmother—began. For three nights we hid nearly four hundred wanted people in a wedding hall—they held a four-day fake wedding to throw off the Taliban. Then, after one extremely tense bus ride to the airport in Mazar-i-Sharif, all twelve of Salam's family members boarded a chartered airliner, alongside hundreds of their fellow "wedding guests."

As the plane taxied toward the runway, Rahim called me. "My brother," he said, "you and your friends are the real mission impossible team, and you have saved us all." I watched on flight radar as the plane took off, and when I saw it leave Afghan airspace, I was flooded with tears of relief.

I sent Salam a two-word text message: "Wheels up."

After the success of Operation Bella, my military friends, my political friends, and I decided we shouldn't stop at one evacuation, so we established the nonprofit Afghan Rescue Project. In the months that followed we helped over a thousand of our Afghan allies escape Taliban-controlled Afghanistan.

But that's a story for another day.

EPILOGUE: THE QUESTION I HAVEN'T ANSWERED

The American Psychological Association didn't establish post-traumatic stress disorder as a diagnosable condition until 1980, but it has been present in human society from the earliest times. As long as there has been trauma, there has been PTSD. Like the devil, it has gone by many names: shell shock, battle fatigue, combat heart. The symptoms of PTSD show up in ancient Mesopotamian texts dating back three thousand years; in them, warriors describe being visited by the ghosts of the people they had killed in battle. In the Mesopotamian epic *Gilgamesh*, the hero witnesses his best friend's death and is driven into a panic that ultimately transforms him completely. The condition appears in the ancient Greek histories of Herodotus, in the Icelandic sagas, and in Indian epics. A fourteenth-century French author described a former soldier who had to sleep far from his wife and children because he sometimes jumped out of bed in the middle of the night and slashed at shadow enemies with his sword.

As long as human civilizations have made war, they have recognized its impact on those who fight it. It's why some ancient cultures held "return rituals" for soldiers when they came home. Some Native Americans held elaborate purification ceremonies in sweat lodges where warriors could tell their stories, watch them catch in the steam rising from the hot stones and dissolve. These fighters had the opportunity to tell their stories aloud, and the whole community could hear and try to understand what they had gone through.

All I did, on the other hand, was get on a plane from Doha.

Despite the fact that we have forty years' worth of psychological research on the prevalence of PTSD in soldiers, that's all anyone in the US military gets at the end of their service: a firm handshake and a goodbye. There was no period of decompression, reorientation, or reintegration.

Native Americans made sure everyone in the tribe understood what the warriors had experienced so that the warriors would never have to feel alone in the tribe. In America today, we give our warriors a free chicken fajita rollup at Applebee's on Veterans Day and expect them to be exactly who they were before the war.

Today only 0.4 percent of Americans are serving in the armed forces, and many veterans come back to find they have no one to talk to who can meet them where they are. As a result they feel isolated. This was not the experience of my grandfather Pop. When he came home, he was surrounded by people who had lived through something similar, such as his brother John, his best friends Bud and Bert, and so many other men of his generation. He never lacked for friends who could talk about it with him. You often hear people say things like "My grandfather never talked about the war." This implies that Grandpa didn't like talking about it; in reality, Grandpa didn't like talking about it *with you*. He didn't think you'd understand. Once I came home from Afghanistan, Pop started talking to me about his time in the military.

It's not that civilians don't want to hear about experiences of war (though often they don't)—it's more like they're not equipped to respond, and they feel bad about that. If a vet says something like "One time my driver took a new route and I almost shot him in the head," what is a civilian—even a close friend—supposed to do with that information? Compare it to the time they had to one-star an Uber driver for driving too fast? People want to connect, but they're scared to get it wrong. We have no equivalent of a sweat lodge where soldiers can share their experiences and where the community can learn to listen with an open mind, without judgment. And vets learn to just bottle it up, telling themselves that what they did was nothing special. That may be the most insidious lie of all.

Yet the national conversation around veterans and PTSD has come a long way. A steady drumbeat of messaging tells them *It's not weak to get help, it's strong.* But that solves only half of the problem.

From the moment that soldiers step off the bus at basic training and up to the moment they render or return their final salute, they receive the clear message that what they're doing is nothing compared to what someone else is doing. That's not a bad thing. In fact it's appropriate, because the jobs soldiers do are hard and frightening. And if we kept that thought in our mind, we might not be able to do those jobs. Basically, the army wires every soldier to think "What I did was no big deal." It's an absolutely necessary form of brainwashing that helps you keep going.

It was frightening to go to meetings over and over again with people who might be luring me into a trap, but I was able to do it because I knew Todd and Kevin (from the tactical intel team) were doing it too, and that made it feel normal. And I could do it because I had been taught it wasn't any big deal compared to what other people must be doing. If, instead, I'd bitten from the apple of knowledge, if I'd had any idea that what I was doing was so dangerous that the job would be completely revamped after Todd and Kevin and I left, I might not have been able to make my legs carry me into the next meeting. If enough intelligence officers had had that reaction, the military wouldn't have gotten the information it needed.

Likewise, if people like Stephen weren't fully convinced that the next squad over or the next platoon over (or even the soldiers from some war of the past) had it much worse, they couldn't go back out on patrol again.

So we shouldn't congratulate ourselves too much for simply re-branding PTSD treatment as an act of strength. Vets will always encourage other vets to get help. We just don't think *we* earned it, because no one ever deprogrammed us from thinking that what we did was no big deal. There is no institutional mechanism—none—for helping service members learn how to flip *off* that switch before they leave the military. No one sits you down and says, "Okay, now that it's over, you should know that what you went through was actually some crazy shit, and you're probably going to need some help."

Or, as Stephen once told me after hearing me compare my service unfavorably to his for the dozenth time: "Man, somewhere right now there's a World War II vet sitting in a VFW hall saying, 'Yeah, I was first wave at D-Day, but no big deal—I was in the back of the landing craft.'"

"No matter who you are," he said, "there's always somebody who you think had it worse. If not for that belief, we'd never win any wars."

This failure to flip the switch back to normal, to deprogram military thinking that is useful during service but potentially harmful afterward, is the reason why I and so many others kept thinking that we couldn't possibly have "earned" our PTSD and that getting treatment for it was like "stealing valor."

That's why it's so important to normalize mental health struggles, and not just for those who have served. No matter who you are, telling your story could save lives.

My decision to go public about my mental health came with certain challenges, but it definitely helped me heal. Not having to put on a show for myself or others made a big difference. I encourage everyone to be up front about their own struggles, in their own way. You don't need a huge platform. Just letting some coworkers or your circle of friends know what you're going through can have strong positive effects, and you'll be glad you did it. In this age of social media, we all live a public life to one degree or another, and you never know who might see your story, follow your example, and give themselves permission to seek help.

And I haven't just healed mentally—I'm living proof that mental health and physical health are inextricably linked.

All those years of disliking myself, of doing nothing about my stress and living on the road, had taken a toll on my body. I had chronic back pain, got sick a lot, and generally felt weak. But once I started to find success in therapy, some of the physical stress stored in my body dissipated, and I found myself able to work out; that in turn made me feel better mentally. When I paired these activities with a commitment to good nutrition, it was a game changer. I went from icing my back and popping ibuprofen multiple times a day to training six days a week for the Annual Memorial Day Murph Challenge, more commonly known as "The Murph."

The event raises money for a foundation that provides scholarships to the children of fallen warriors. It's named after the late Lieutenant Michael P. "Murph" Murphy. A navy SEAL, Murph was posthumously awarded the Medal of Honor for his heroic efforts to save his team from a Taliban onslaught about fifty miles north of Jalalabad in 2005. I had known about The Murph for a while, but never in my wildest fantasies did I imagine actually attempting it; that seemed insane.* But once fitness became a full-blown hobby of mine—I had never had a hobby before—I surprised myself and completed The Murph. And with a pretty decent time to boot.

Now I do The Murph every Memorial Day. It takes me over half an hour, and I spend every sweaty, excruciating minute—every grueling repetition—thinking about people who didn't come back whole, guys like T.J., as well as people who came back but still haven't made it home. It has become more than a fitness challenge. It is my own personal return ritual. My annual purification ceremony, my penance. It reminds me that I was—and am, and always will be—part of something greater.

Getting more involved with Veterans Community Project also gave me that good connected feeling, but on a different level. A few months after Bryan Meyer and the VCP team helped me cut through the VA bureaucracy, I just started hanging around the place. In theory, I was mentoring the founders as they figured out how to say yes to the invitations they'd received to expand into communities across America—for an organization like VCP, which relies on

---

* A one-mile run, one hundred pull-ups, two hundred push-ups, three hundred air squats, and then another one-mile run to finish. All of this is done, by the way, while wearing body armor or a weighted vest. Apparently, Lieutenant Murphy designed this workout as an annual challenge to his SEAL team.

having an incredibly sophisticated and intelligent team of employees on site, scaling up was a challenge. But honestly, I just loved the vibe. Nobody cared who I was. The only thing that mattered was whether I could help the mission. I loved everyone's sense of humor, which could be described as pitch-black. After all, the place is made up largely of vets with PTSD—the inmates were running the asylum. But somehow, they were actually helping everyone get better. It felt like home.

Then one day, Bryan popped the question: "Hey man, we're working to become a national presence, and you've built a national operation before. So instead of teaching us how to do it, how about coming on board full-time?"

Today I'm the president of national expansion at Veterans Community Project, and we're serving veterans and building campuses across America. All the combat vets leading VCP, including me, know that if just a few variables in our lives had played out differently, we'd be living in one of the tiny houses instead of working at national headquarters. Like everyone I work with, I wear a VCP T-shirt or hoodie pretty much every day, and it feels good to rock that uniform.

Both as secretary of state and as a prospective presidential candidate, I'd taken dozens of tours of nonprofit organizations. Back then, I loved showing up in the black SUV with a small entourage, meeting the people who were nervously awaiting my arrival, walking through, and asking questions with a throng of reporters in tow. It perfectly fed my need to feel that I was doing important things, my need for external validation. It gave a boost to that starving internal glimmer of hope, whispering to me that I wasn't irredeemable.

When my team drove off in our imposing chariot, heading to the next event, I sometimes thought about the people staying behind, the staff or volunteers. It filled me with a mix of pity and envy. I pitied them for living "small" lives that didn't come with an entourage of handlers in a big black SUV, but I envied the fact that their work held unmistakable meaning. And I could see that they were happy.

Over the past couple of years, I've given tours of VCP to governors, senators, and presidential candidates. Afterward, I watch them climb back into a big black SUV and ride away to the next stop on a packed calendar, making fundraising calls on the way. Then, with a smile, I walk back into our national headquarters and hang out with my friends. Though I don't feel envy, I sometimes do feel sad for those dignitaries. Come evening, they won't get to have dinner with their families.

True remembers that Daddy used to ride on airplanes for work and that Let America Vote had thundersticks, but nothing about those chapters of my life makes him proud of me. If you ask True about those days, he'll tell you that sometimes he forgot that he had a daddy at all.

Now he adores coming to work with me at VCP. He helps the staff and loves meeting new residents of the village. One day this past winter, as True and I were waiting at a stoplight and I was staring out the windshield, my mind wandering, True asked me a question from his booster seat in the back: "Daddy, do we have a dollar we can give that man?"

I looked to my left and noticed the homeless man, holding a sign, on the median strip a few feet away.

"Not sure, buddy. Let me look," I said as I opened the center console.

I found a single, rolled down my window, and gave it to the man in the old drab-olive coat just as the light changed.

I wanted to offer some positive reinforcement for True's generosity, so I said, "Thanks, buddy. I wouldn't have noticed him if you hadn't said something. What made you want to give him a dollar?"

"It's going to be cold tonight, and he looked like he doesn't have a house and he had on an army jacket. I think he might be an army man, and helping army men who don't have houses is what you do, Daddy."

I held back the tears. "It sure is, buddy."

In 2006 I put my body on the line because I believed that I could help a few more Americans get home safely, but I never felt as though I'd accomplished it. In 2018, in the days after my announcement became a major national news story, I learned that calls to the VA Crisis Line had tripled.* When Diana told me that, I became so emotional, I could barely speak, but I did manage to utter these words: "This is the first time I've ever felt as though I might have helped someone get home safely." Today at VCP, I get to be a part of accomplishing that mission every day. We're helping vets come home again.

But there are still untold thousands out there who aren't getting the help they need.

A few months ago, after fourteen years of thinking about him, I got back in touch with Todd (from the tactical human intelligence team), who was still on active duty. Todd told me Kevin fell apart even before his tour ended, and he had to be sent home. Within a year Kevin died in that one-vehicle accident. A few years later, Todd survived his own one-vehicle accident. Apparently, back in 2007, after I rotated home, Todd tried in vain to replace me, but no candidate for the job made it past their audition. According to Todd, one guy—a reservist who was a cop in civilian life—literally urinated on himself during his tryout. Todd went back to Kabul for another tour, and by then our jobs were being done in a completely different way because, and I quote, "What we did was fucking insane, Jason."

For the past fourteen years, Todd had been having the same symptoms I had, and he promised me he'd make treatment a priority in a few months, when he retired from active duty. I didn't push him to get help immediately—I figured if he'd made it this far, he could last a few more months.

But a few weeks later, Todd downed several handfuls of pills and pulled his truck onto the side of an abandoned road. As he drifted into unconsciousness, he was confident the pain was coming to an end. Minutes later, a sheriff's deputy happened upon his truck. When Todd woke up in the hospital five days later, he was pissed to learn he wasn't dead.

He spent several weeks in an in-patient treatment facility at the

---

* One more time, consider putting it in your phone: 1-800-273-8255.

VA, and he soon decided that he didn't want to die after all. At some point a staff member showed Todd a video of another combat vet talking about post-traumatic growth—and suddenly there I was, his old buddy from Kabul, on the screen.

These days, Todd and I stay in touch. After treatment, at first he lived alone in a fifth-wheel camper because his marriage had ended, but he found his way through the divorce and back into his own house. He got a therapy dog, and he's rebuilding his relationships with his kids. He goes to therapy and he does his homework. Todd, like me, is evangelical about post-traumatic growth, and he plans to do a lot of suicide awareness work in the future. I'm proud of him.

But what if Todd and Kevin and I had stayed in touch when we'd come home? What if we'd found out we were having the same nightmares, being hypervigilant, and feeling convinced we hadn't done enough in Afghanistan? Would Kevin still be alive? Would Todd and I (and our families) have been spared years of difficulty? I suspect so.

Being the poster child for post-traumatic growth is not the role I envisioned for myself when I entered public service, but as long as I have a platform and influence, I want to make the most of it and carry out this mission. I want to spend the rest of my life catching all the Todds and Kevins I can before they fall too far.

Which, I guess, brings us to the Big Question, the one I get asked pretty much every single day: "Are you ever coming back?"

I can't tell you how many times a day I see messages telling me what I have to do—I have to be the Stacey Abrams of Missouri, I have to run against whichever proto-fascist clone is trending on Twitter at that moment. And honestly, sometimes I'd like to respond, "No, I don't. I don't *have* to do a damn thing. *You* do it."

Don't get me wrong—it is cool to be asked. And I get it. I'm the Natural, so I'm *supposed* to want this. When someone asks this question in an interview, usually the muscle memory from when I pretended I wasn't running for president kicks in, and I say, "I'm just focused on helping veterans right now." It's a politician's trick: tell the truth, but not all the truth.

But I've had more than enough hiding and bullshit for one life-time. If you've read this far, you deserve the real answer, and here it is: I genuinely do not know.

Sorry.

I'm probably "coming back" one day—that is, in the way most people imagine: running for office. But not now, and not for a long time. Not until I know I can make a difference every single day as a politician *and* enjoy my life. When I think about whether to take a new job, I ask myself whether I want it more than the job I've already got. Right now, if someone could wave a magic wand and make me a US senator, I'd say no, don't do it—because the job I have today is just flat-out better.

After I dropped out of the mayor's race, my calendar suddenly emptied, a white tundra of nothing to do. But one appointment had been set to repeat indefinitely: the Monday afternoon scheduling call that broke my heart every week. For some reason, I didn't delete it then. Every Monday at 2:30 p.m., my phone pings to notify me that I have a scheduling call—only I don't. I leave it there to remind me of everything I never want to go back to.

And one of those things is the unmanageable load of guilt I attempted to carry for so long. I put myself in that "lose/lose" mindset so that if anything good happened, I would tell myself I didn't deserve it or that someone else deserved it much more than I did.

But not anymore. For example, I get a check every month from the VA—and I keep it. Here's why: A VA diagnosis of service-connected post-traumatic stress disorder is required in order to be treated at the VA's PTSD clinic, and with that comes a VA disability rating. And with a VA disability rating comes a monthly VA disability check. At first, I was deeply uncomfortable about this, in part because I was still comparing my trauma to others' and struggling with feelings that I "hadn't done enough." But once I worked through those feelings, I still felt self-conscious about receiving a government check in the mail each month.

I wasn't working at the time, but I knew my earning capacity was high, and Diana's income was enough to support us both. In short, we were doing well, and it felt wrong that I was receiving this monthly check.

I brought this up with a benefits specialist at the VA, and he told me about a veteran who'd come in for mental health treatment for service-connected PTSD after retiring from a very successful career as an entrepreneur. Once he got to know the guy well, he asked him how he felt about receiving this check each month, which for him amounted to little more than a drop in the bucket.

"Thirty years ago, I gave the air force a perfectly healthy young man," the man said. "And two years later, they gave me back a deeply troubled old bastard. So the way I figure it, that check is the least they can do." That story resonated with me.

I now understand that I've actually given quite a lot of myself. I still do things for my country, and I still engage in politics, but no longer out of a sense of obligation, as if I still hadn't done enough. Now I do it because I enjoy it, or because I want to advance a cause I care about, or because it makes me feel useful, and for a variety of other reasons. But I rarely do something because I think I *should*.

I'm still out here trying to better this country, but it's because I *want* to, not because I feel I *have* to.

Today, America and I are square.

But not running for office doesn't mean I'm not in the fight. Being "in politics" isn't just about putting your name on a ballot—ask the House aide who spends years anonymously grinding for a single policy she genuinely believes in.

I relaunched my podcast, *Majority 54*, which remains one of the top political pods in the country. Let America Vote merged with another group to form the largest democracy-focused political organization in America, and I serve on the board of directors.

As weird as it sounds, I'm a forty-year-old party elder. During the 2020 campaign, I advised most of the Democratic candidates for president on veterans and national security policy and—at the urging of Tom Perez, who was then the chair—I served on the DNC platform committee. I headlined virtual events for progressive candidates and causes at least four evenings a week and was a Biden surrogate tasked with being the Red State Whisperer on Fox News. The week of the election, I got a call from people on the Biden-Harris transition team. They mentioned a couple of possible cabinet positions and wanted to know if I was interested in being considered. Diana and I daydreamed about it for a couple of days before politely declining. That surprised a few people, but the minute I step back on that treadmill, I have to give up everything else, and that's just not what I want. Not right now.

My calendar is full again, but unlike before, it's not all work. I take True to school almost every morning, and I'm there to pick him up almost every afternoon.

I even joined a baseball team again. I play in an over-thirty wood bat league, and my parents come to all my games. True and Diana often join them, and—let me tell you—thousands of people chanting your name at a political rally pales in comparison to hearing your son yell "Yeah!" after you steal a base with a head-first slide.

And I drive a pickup truck, which is loaded down with baseball equipment because I coach True's Little League team. Just like my dad coached mine, and just like Pop coached his.

Today the real powerhouse in the family is Diana. She's an author, a podcaster, a highly sought-after consultant, speaker, and thought leader. She's an incredible mother and the spiritual and emotional rock of our family. Recently, at her suggestion, we started offering prayers of gratitude before meals, like I did in Afghanistan.

We're having fun together again. She's even training to do The Murph with me next year. And she finally taught Talia how to cuddle on the couch. Talia's a big softie now. But if you ever meet her, it's best not to make direct eye contact.

DIANA

I don't feel like a powerhouse. I feel like I'm finally free to explore the things I want to work on, and it feels good. I'm also free to say no to things, and saying no has become an essential practice for both Jason and me.

We've made an annual family holiday out of October 2—the day Jason announced that he had PTSD and withdrew from the mayor's race. All of us find something in our lives that hasn't been serving us, and we say no to it. True still hasn't quite gotten the hang of it—he keeps trying to say no to eating eggs.

Just like Jason, I try to be as open as possible about my own mental health struggles in an effort to help others seek help. The thousands of letters Jason received in response to his announcement and the many people who confided their personal struggles to us because they thought we wouldn't judge—it gave me perspective. Many people are living lives of quiet desperation, alone and ashamed, just like Jason and me for over a decade.

I feel like I'm on my second marriage because so much has changed between us. When I think about the man I'm married to, a caring and thoughtful husband and father who is the epitome of a partner, it is an overwhelming feeling. We still have bad days, we still have arguments, but mostly our life is filled with the fun of being together and making an impact. It's what we dreamed of at seventeen.

Meanwhile at VCP, I use the skills I honed in politics every day. I persuade local leaders to let us build our villages in their neighborhoods, use my social media following and my contacts in traditional media to recruit volunteers, and set the vision for our national expansion. And because I knew nothing about construction or social work two years ago, I'm constantly challenged to learn as well.

Those celebrity friends I was worried about losing? By and large, they stuck around, and now I get to harass them constantly about using *their* platforms to promote causes and campaigns I care about, like VCP or Afghan Rescue Project.

I still think I would be a hell of a president. And I still love my hometown and might choose one day to go for mayor of Kansas City. And someday, when I don't have Little League to coach and homework to help with, when the people in my house are good and bored of me, I may be ready to do that. But not yet.

Or maybe I'll spend the rest of my career putting VCP campuses in every city in this country until no veteran falls through the cracks and every homeless veteran has been housed. That would suit me just fine, because, as True says, that's what I do.

I spent years strategizing—in great detail—what I *planned* to do in the future because thinking obsessively about the future was preferable to living in the intolerable present. But now I'm enjoying the present. I'm done doing things so that later I can have the chance to do other things. I serve my country by doing things that matter right now.

Remember that book I was reading the day I walked into the VA feeling suicidal? Let me tell you about the guy who wrote it. People called Rick Ankiel "the Natural" too. Pitching seemed to come as easily to him as breathing. From the very first year he was called up to the St. Louis Cardinals, people pegged him as a future first-ballot Hall of Famer. At the end of that rookie season, in the third inning of game one of the 2000 NLDS, he walked four batters and threw five wild pitches. It was as if something had come unstuck in his brain. Out of nowhere, he had developed the yips—a death sentence for athletes. Baseball players with the yips suddenly find themselves launching pitches to the backstop. No one knows where the yips come from or how to get rid of them, but they're thought to be neurological in origin.

The next season, it got worse. Frantic, Ankiel tried everything possible to cure himself. Finally he settled on getting ripped on vodka before he pitched. When the vodka didn't work, Rick Ankiel's major league career seemed to be dead. "The Natural" languished in the minors for the next six years—in sports, the equivalent of a lifetime. And then in 2007, the Cardinals called him back up. Why? Because he had given up doing the thing he had once done better than almost anyone else alive, the thing that had defined his entire life up to that point: pitching.

Rick Ankiel became an outfielder. He was never great again, but he was good enough to stay in the majors until he retired in 2014. (And he had a knack for gunning down base runners from deep center field.) He was able to continue playing the game he loved while also enjoying his life and his family.

For now, at least, that's all I want. I'm giving up being a pitcher for a while, and I'm giving the outfield a shot. Almost two years to the day after I admitted to the world that I needed help, Diana gave birth to our daughter. Her name is Bella Brave Kander.

It's a whole new ball game.

AUTHOR'S NOTE

Thank you for reading *Invisible Storm*. I'd love to hear your thoughts. Please post a picture of the book in the place you were sitting when you finished reading it, then tag someone you think would enjoy reading it too. And don't forget to tag me (@JasonKander on Twitter and Instagram) so I can thank you personally!

ACKNOWLEDGMENTS

Invisible Storm would not have happened without Samuel Ashworth. I knew I wanted to write this book, but unlike with my first book, it was tough to open a blank document and fly solo into a story I hadn't always enjoyed living. I needed someone who could help me sort out which experiences to include and how to articulate the accompanying emotions. And that meant I needed someone who knew me, cared about me, and who also happened to be an extraordinary storyteller. Enter novelist, journalist, and teacher Samuel Ashworth. Sam is a brilliant, award-winning writer who also happens to be my first cousin. He was my coach and my collaborator from start to finish, and there's no world in which I could have written this book without him. I'm eternally grateful.

My wife Diana never settles for "good enough" in anything she does, and I'm grateful that she brought her magical talent to *Invisible Storm*. Reviewing drafts and composing and inserting her own thoughts was an emotional and often arduous task, but she did so much more than that. The first draft was ten thousand words too long, and I have a tendency to wed myself to stories and riffs I've composed. Diana took the time to find those ten thousand beautiful but unneeded words and—with patience and love—convince me to set them free. She did this while simultaneously supporting me through whole weeks when I had to write particularly upsetting passages. She made the book better in every way.

Mel Berger, my wise and capable agent at William Morris Endeavor, gave me confidence by expressing his belief that I was writing a necessary book. Rakia Clark, my editor at Mariner, brought excellent perspective and focus to the text. Writers can be sensitive about editing, but her friendly constructive criticism made the process a pleasure.

I had a great experience with Mariner books and HarperCollins. From Brian Moore's smart jacket design to Susanna Brougham's attentive copyediting, Ivy Givens's organizational prowess, and Lisa Glover's careful management of production, it was a top-notch ride.

Because I was at a loss to come up with a book title, I asked the Twitterverse for ideas. I'm thankful to Sara Flick of Minnesota for replying with *Invisible Storm*.

Several people read drafts and offered suggestions that improved the final product. I want to thank Janet Kander, Steve Kander, Abe Rakov, Stephen Webber, Kellyn Sloan, John Kander, Albert Stephenson, Susan Kander, Warren Ashworth, Bryan Meyer, Ravi Gupta, Jack McCracken, Todd Schoerberl, Stacey Abrams, Will Kanatzar, Jason Sudeikis, Grace Lynch, Sarah Whitten, Chloe Hall, Candy Smith, and Nicholas Heinecke.

The team members at Veterans Community Project are always patient with me, but they were particularly great about giving me time to work on this book. If you're interested in supporting our efforts, I'd encourage you to check out VCP.org.

Finally, I want to thank all of my family and my friends—not just for their support as I wrote this book, but for all the years they stuck by me when I wasn't quite me.

Most of all, thank you to Diana, True, and Bella. For everything, especially the family hugs. I love y'all more than anything in the world.